Studies in Systems, Decision and Control

Volume 374

The series "Studies in Systems, Decision and Control" (SSDC) covers both new developments and advances, as well as the state of the art, in the various areas of broadly perceived systems, decision making and control–quickly, up to date and with a high quality. The intent is to cover the theory, applications, and perspectives on the state of the art and future developments relevant to systems, decision making, control, complex processes and related areas, as embedded in the fields of engineering, computer science, physics, economics, social and life sciences, as well as the paradigms and methodologies behind them. The series contains monographs, textbooks, lecture notes and edited volumes in systems, decision making and control spanning the areas of Cyber-Physical Systems, Autonomous Systems, Sensor Networks, Control Systems, Energy Systems, Automotive Systems, Biological Systems, Vehicular Networking and Connected Vehicles, Aerospace Systems, Automation, Manufacturing, Smart Grids, Nonlinear Systems, Power Systems, Robotics, Social Systems, Economic Systems and other. Of particular value to both the contributors and the readership are the short publication timeframe and the world-wide distribution and exposure which enable both a wide and rapid dissemination of research output.

Indexed by SCOPUS, DBLP, WTI Frankfurt eG, zbMATH, SCImago.

All books published in the series are submitted for consideration in Web of Science.

More information about this series at http://www.springer.com/series/13304

Zdzislaw Trzaska

Mixed Mode Oscillations (MMOs)

Fundamentals and Applications

Springer

Zdzislaw Trzaska
Warsaw University of Technology
Warsaw, Poland

ISSN 2198-4182 ISSN 2198-4190 (electronic)
Studies in Systems, Decision and Control
ISBN 978-3-030-76866-9 ISBN 978-3-030-76867-6 (eBook)
https://doi.org/10.1007/978-3-030-76867-6

This Springer imprint is published by the registered company Springer Nature Switzerland AG
The registered company address is: Gewerbestrasse 11, 6330 Cham, Switzerland

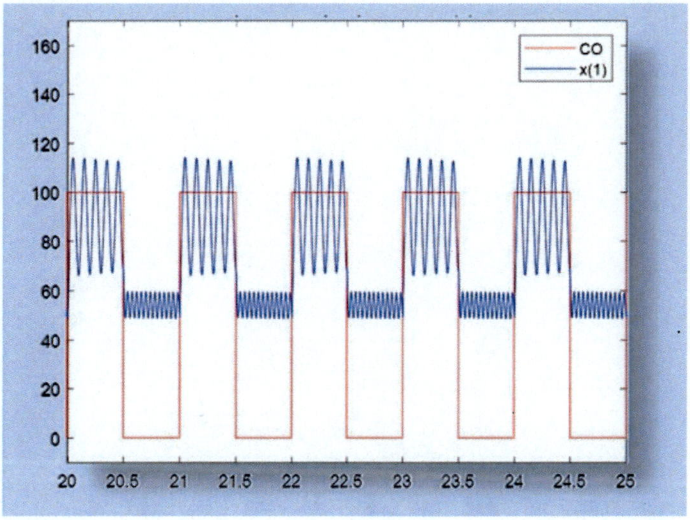

CO
x(1)
160
140
120
100
80
60
40
20
0
20
20.5
21
21.5
22
22.5
23
23.5
24
24.5
25

Preface

This book focuses on plenty of problems relating to mixed mode oscillations (MMOs for short) that often occur in real nonlinear dynamical systems. The problem of dissection of MMOs into a sequence of small-amplitude oscillations (in brief SAOs) and of large-amplitude oscillations (in brief LAOs) is highlighted. A typical sequence of transitions is that even a relatively uncomplicated system, described in terms of deterministic evolution equations, may exhibit a sudden change to a completely new, qualitatively different behavior after smooth changes in the control parameter. Mastering this art guarantees the correctness of the procedures used to study nonlinear dynamical systems. In this book, we strive to meet such challenges.

The emphasis is on existing and systematically emerging new methods in the field for scientists, engineers and practitioners working in many, often interdisciplinary, fields where nonlinear dynamics plays a key role. These areas include analytical and environmental sciences, control engineering, electrical circuit analysis, mechanical systems and biochemical and medical research as well as in economics, among others. This is the first comprehensive approach to this topic emphasizing the nature of the phenomena studied and the effects of their analysis. Properties of MMOs are illustrated by several examples presented in each chapter.

Each chapter of the book is, to a large extend, self-contained with its own notation and method of presentation. Coverage includes:

- Dynamic behavior of nonlinear systems
- Fundamentals of processes exhibiting MMOs
- Mechanism and function of a structure of MMOs' patterns
- Analysis of MMOs in electric circuits and systems
- MMOs in chemistry, biology and medicine
- MMOs in mechanics and transport vehicles
- MMOs in fractional-order systems.

This is the first complete introduction to the mixed mode oscillations' identification and control that fully combines software tools—providing professionals and students to hands-on master critical techniques, thanks to computer simulations based on the straightforward MATLAB environment.

The center of concern is set on existing as well as emerging continuous methods for researchers, engineers and practitioners active in the many and often interdisciplinary fields, where electrochemistry, biology and medicine play a key role. These fields will range—among others—from analytical and environmental sciences to sensors, materials sciences and biochemical research. This is the first extensive description of this topic and very demanding physically in-depth insight in the interpretation of analytical results and those obtained from computer calculations. One should emphasize that the mechanisms and conditions for the formation of spatial and spatiotemporal patterns are of tremendous importance for understanding of such phenomena in chemical, cardiovascular and biological systems.

This book falls into a category intended to survey MMOs in practical systems occurring in many applications, such as electrical and electronic engineering (circuits and control systems), mechanical systems and transport vehicles as well as in electrochemistry and biological structures. These applicable areas proved to be filled with many interesting MMOs in various structures and forms. Reading the book can provide researchers with better and better knowledge not only to carry out a qualitative analysis and calculations of a given model, but also to independently identify the phenomena underlying the formal and mathematical description of the system under study.

The material is presented in seven chapters, and each of them is divided into sections. Usually, the first section of a chapter is of an introductory nature, explains phenomena and exhibits numerical results. Investigations of a more advanced nature are presented in the subsequent sections of each chapter. The bibliography contains only those papers that are referenced in the text and is no way meant to be complete.

It is hoped that the book will provide the readers with better understanding of the nature of MMOs, richness of their behaviors and interesting applications.

Warsaw, Poland
April 2021

Zdzislaw Trzaska

Contents

Chapter 1
Preliminaries

1.1 Cornerstone

Everywhere in nature occur oscillators producing rhythms and vibrations in forms of a strong, regular, repeating course of motion, lightning, or sound. Simple oscillators are based on a single mechanism, such as in the case of a usual pendulum without damping performing harmonic oscillations, but in more complex systems different processes occur at various phases of the oscillation. However, real world oscillators rarely exhibit the uniformity of the harmonic oscillator but typically alternate between slow and fast course with large and small amplitudes, respectively. Such mixed mode oscillations (abbreviated as MMOs) are the subject of a considerable current research effort, and include experimental, computational, and theoretical treatments which guide light on important issues in physics, chemistry, engineering, biology, medicine and economy. The research area of MMOs is a relatively new one, but their traces and algebraic foundations can be found early (more than 100 years ago) in the works of B. P. Belousov and A. M. Zhabotinsky on nonlinear chemical reactions [1]. The oldest examples of analogous phenomena include oscillations of electric current upon anodic electro-dissolution of some metals. Periodic dissolution of metals was described as early as in 1828 by G.T. Fechner, who reported repetitive bursts of effervescence (gas bubbles evolution) during the dissolution of iron in nitric acid [2]. Roughly speaking, MMOs are the time series of a dynamical system in which there is an alternation between oscillations of distinct large and small amplitudes, abbreviated as LAOs and SAOs, respectively. There is no widely accepted strict criterion for this distinction between amplitudes, but usually the graduation between large and small is clear (e.g. in Fig. 1.1). In slow-fast systems the SAOs arise in a localized region of the phase space and, on the other hand, the LAOs are connected with large deviations away from the concentrated region of SAOs.

The explosion of interest in MMOs began at the turn of the twentieth and twenty-first centuries among the community of chemists, biologists, cardiologists, neurologists, engineers and mathematicians. Such specialists, working jointly, try to explain as many aspects and effects of MMOs as possible. How they achieve this holds

Z. Trzaska, *Mixed Mode Oscillations (MMOs)*, Studies in Systems, Decision and Control 374, https://doi.org/10.1007/978-3-030-76867-6_1

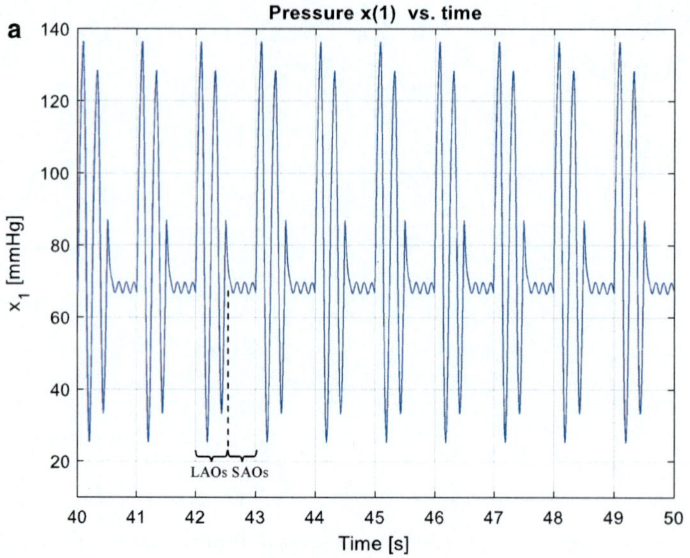

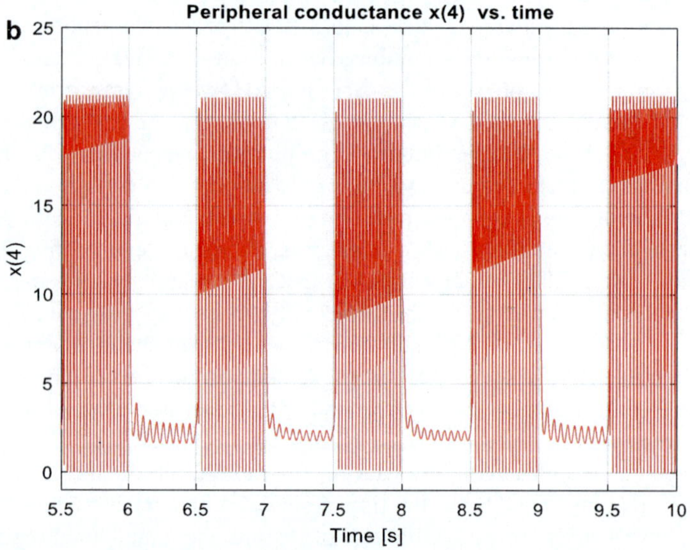

Fig. 1.1 MMOs at the load of a human left ventricle: **a** blood pressure variations at input of aorta, **b** variations of peripheral conductance in the cardiovascular system

motive lessons for the way in which various disciplines—not only those in the physical and biological sciences—could more successfully engage with each other on common problems. This leads presently to main ideas concerning fundamentals and applications of MMOs.

Substantial recent research efforts, including experimental, computational, and theoretical approaches shed light on important MMOs issues in physics, chemistry, and biology as well as from numerous other disciplines of human activities. They are focused on MMOs observed in, among others, the tunnel diode and Bonhoeffer-van der Pol oscillators, the Hodgkin-Huxley model of neuron dynamics, Taylor-Couette flow, surface oxidation and autocatalytic reactions, human heart arrhythmias and in epileptic brain neuron activity [3]. MMOs with intrinsic subthreshold (membrane potential) oscillations (STOs) at theta (4–12 Hz) and gamma (30–100 Hz) frequency bands have been observed in various neuron types and have been implicated in cognitive processes including memory, spatial navigation, and sleep [4, 5].

Typically, mixed-mode oscillations are complex periodic waveforms comprising a certain number of large and small amplitude excursions in some particular order. These amplitude regimes differ roughly by an order of magnitude. In each regime, oscillations are created by a different mechanism and their amplitudes may have small variations in time. The transition among regimes are governed by additional mechanisms. Generally, the term "mixed mode oscillations" means the spontaneous formation of the set of ordered values in the time domain when the system is maintained sufficiently far from equilibrium. There are two reasons for this condition: the thermodynamic and kinetic one. The first one is justified by the second law of thermodynamics: the creation of any order, including dynamic self-organization phenomena, is associated with the decrease in entropy. Therefore, there must exist, in the same system, a dissipative process characterized with entropy production at least compensating this decrease. Increases in entropy correspond to an increase of disorder and to irreversible changes in a system. Entropy measures the degree to which energy is mixed up inside a system, that is, the degree to which energy is spread or shared among the components of a system. Entropy is large when the small microscopic components of the system are disorganized and move completely independently. Every physical system, with the passage of time from its origin, spontaneously approaches a state of equilibrium. In a state of thermodynamic equilibrium, there are no net flows of matter or of energy, no phase changes and no unbalanced potentials (or driving forces), within the system. The change in entropy in a dynamical system is always higher or equal to zero, and time is the fundamental dimension in which the system is doing work. When energy is dispersed or widely distributed in a system, the possibility to use that energy for mechanical work decreases and entropy rises. From the kinetic point of view the MMOs may take place in larger intervals as the dynamic switches between SAOs and LAOs. They can be subsequently identified in a number of 2D singularly perturbed oscillators. All such systems are characterized by a presence of a small parameter that makes the systems singularly perturbed. Usually a cubic nonlinearity is typical in systems exhibiting MMOs. Consequently, typical time series of MMOs in such systems consist of oscillatory successions in which amplitudes of different orders of magnitude alternate.

An important kinetic requirement for MMOs generation is that the differential equations describing dynamical systems must be nonlinear, must be in dimensional order at least 3, and span multiple time scales. However, it is not a unique sufficient condition for the occurrence of self-organization in the system, e.g., generation of spontaneous oscillations of the dynamic variables. Moreover, the analysis of the origin of such phenomena also clearly points out at the necessity of the existence of positive and negative feedback loops in the links that make up the entire mechanism of a given process. Therefore, the production of a given manifestation of self-organization requires, among others things, introducing appropriate feedback loops. In fact, even the simplest case of oscillatory proceeding in time can be considered as the interaction of minimum two feedback loops: a positive feedback loop and negative feedback loop. During the activity of the positive feedback loop, the increments of given system's variables growth in time at a great rate up to the moment when the negative feedback loop takes over the control and leads to a decrease in these changes until the instant at which positive feedback sets in again, and the process repeats in time continuously. Hence, the oscillatory process in the dynamics of the system requires the activity of fast positive and slow negative feedback loops. Quite simple conformity can be recognized in the case of oscillations of a plain pendulum when the positive feedback loop is initiated at the lowest position (then the movement's velocity is maximal). However, at the highest position appears the negative feedback loop which firstly annihilates the movement and next changes the velocity direction to the lowest position.

A seminal example is the observation made in [6] with respect to activity of a human cardiovascular system. The time series representing the steady state blood pressure at the output of the left ventricle and the peripheral conductance in the form of MMOs for pulsating heart volume flow are presented in Fig. 1.1. In such a case, the system can be considered as a combination of linear and nonlinear oscillators.

The inter- and intra-spike frequency may belong in the same frequency range or not, depending on the system type and the parameter values. A remarkable number of models for these phenomena lead to differential equations with multiple timescales [5, 7]. Usually, it suffices to consider two timescales and study fast–slow ordinary differential equations (ODEs) which already provide many generic mechanisms producing MMOs [5],

1.2 Concise Characterization of MMOs

As referred briefly in the previous section, the MMOs is one of the most important and most general features of nature, being practically omnipresent in our world, viz., in physical and inorganic systems, in organic and living systems, as in neuroscience, e.g. stellate cells [8], Hodgkin– Huxley-type neurons [9] and pituitary cells [10] and even in social systems as well as in economics [1]. Recall that MMOs occur when a dynamical system switches between fast and slow time-varying changes with small and large amplitude, respectively. The basic idea is that the small-amplitude oscillations (SAOs) are induced by a local mechanism while large-amplitude oscillations (LAOs) are produced by a global return mechanism. Typically, a basic MMOs cycle consists of s small amplitude oscillations (SAOs) followed by L spikes (LAOs) or vice versa. The notation L^s is conventionally used to describe these waveforms [1, 11].

Overall MMOs waveforms can be regular (Fig. 1.1a) or irregular (Fig. 1.1b). Moreover, regular MMOs waveforms can be uniform, where the sequence L^s describes the MMOs waveforms in their totality (Fig. 1.2a), or nonuniform, where MMOs cycles consist of sub-cycles described by sequences of the form $L_1^{s_1} L_2^{s_2} \dots L_n^{s_n}$ (for some integer n) (Fig. 1.2b) called the MMOs signature; it may be periodic or nonperiodic. Signatures of periodic orbits are abbreviated by giving the signature of one period.

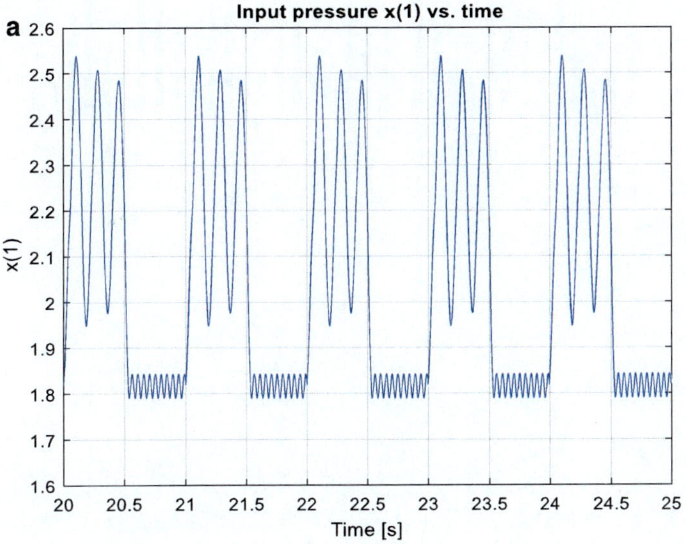

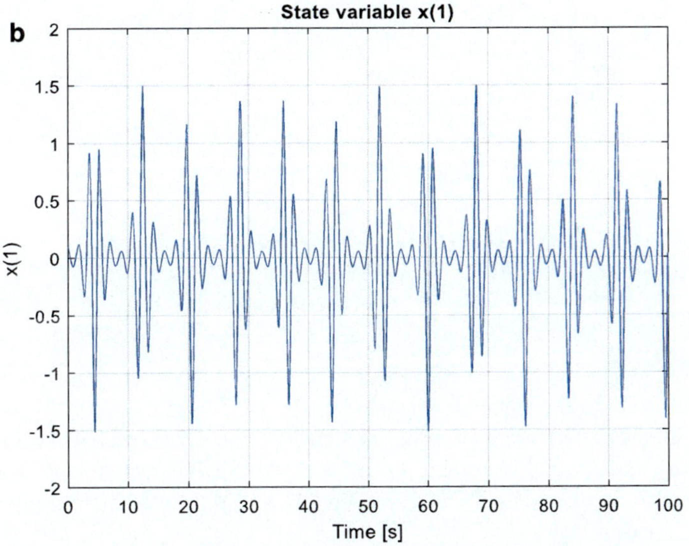

Fig. 1.2 Waveforms of MMOs: **a** regular, **b** irregular, **c** stochastic, **d** chaotic

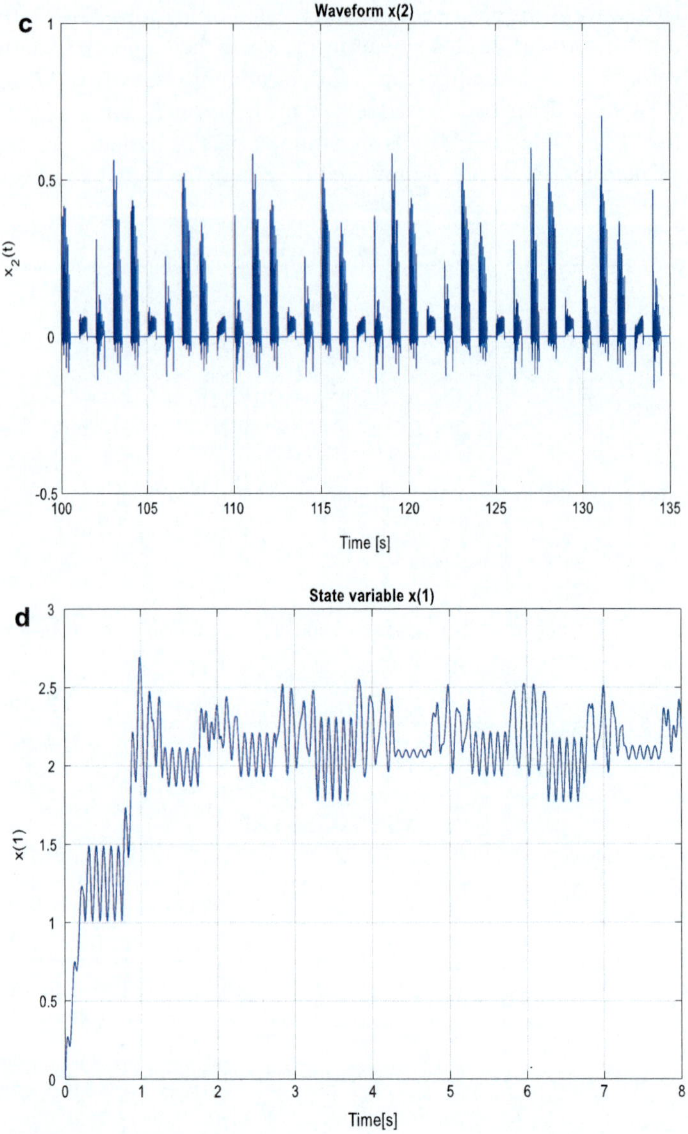

Fig. 1.2 (continued)

Irregular MMOs patterns can be either stochastic (Fig. 1.2c) or chaotic (Fig. 1.2d). The occurrence of L^0 cycles (for $L > 1$) embedded in MMOs waveforms, with two or more consecutive spikes with no interspersed STOs (Fig. 1.3), is customarily referred to as spike clustering [9].

MMOs produced in systems described by three-dimensional ODEs with one fast and two slow state variables exhibit prototypical mechanism in the form of

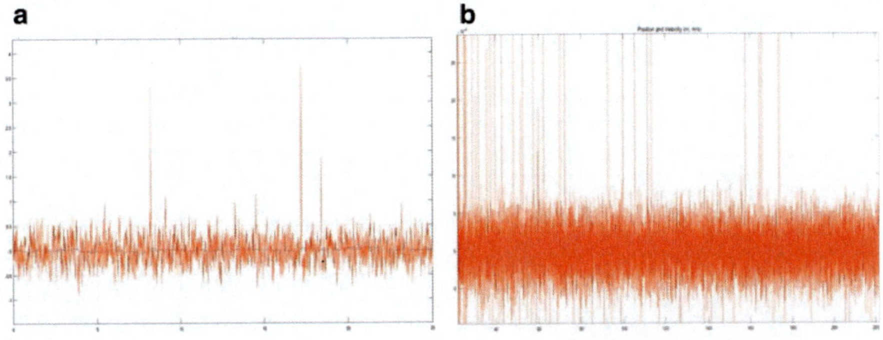

Fig. 1.3 MMOs waveforms with no interspaced spikes: **a** single spikes, **b** multiple spikes

folded-node singularities which are generic for SAOs [12]. The global return mechanism, which is generic for LAOs frequently encounters a relaxation type oscillations induced by a cubic (or S-shaped) fast-variable nullcline, also called the critical manifold. It was studied extensively yet in the case of van der Pol-type oscillators [13, 14].

For a detailed review of the topic see [10, 15, 16], the special issue [5], and references therein. Higher dimensional systems provide favorable flexibility to generate more complex MMOs waveforms. Usually, in these systems, such individual mechanisms as STO and spikes are excited by a different, i.e. lower dimensional regime [9].

Most of the MMOs are analyzed in more details in the subsequent chapters. The chosen examples and topics do not exhaust all possible discussion topics on MMOs and areas of their applications. The new topics will certainly emerge in a near future as MMOs seem to attract more and more mathematicians, biologists, physicians and engineers.

1.3 Mixed Mode Dynamics, Stability and Bifurcation

In general, most real-world phenomena are dynamic, non-linear, and time-varying. Although they appear mainly in the world of physics, it is worth noticing that their dynamics indicates on impressive analogies with other types of dynamical systems, such as chemical, medical, biological, and even appearing in some social processes or market exposure capital.

Such problems as, among others, the bursting dynamics in Taylor–Couette flow, the compartmental Wilson–Callaway model of the dopaminergic neuron, stellate cell dynamics, the famous Hodgkin–Huxley model of neuron dynamics, the coupled calcium oscillator model, the complex plasma instability model, and the surface oxidation reaction and autocatalytic chemical reaction (for such examples see [5, 17]) are described by singularly perturbed systems of nonlinear ordinary differential equations (ODEs) with three (or more) state variables changing at different time scales (usually two or three).

Since each of these categories of processes has to deal with its specific language, then the universality of dynamic behavior is clearly manifested exclusively at the level of their mathematical description, which in turn is based on the preservation of the solutions of the respective nonlinear differential equations. Therefore, the current achievements in the field of dynamics of nonlinear systems constitutes a significant help in the analysis of MMOs occurring in real systems.

The mechanism of generation of MMOs crucially depends on the mechanism of transition from STOs to spikes. Several mechanisms have been described in the literature [12]. Among the various mechanisms which explain the occurrence of MMOs are the break-up of an invariant torus [18] and the loss of stability of a Shilnikov homoclinic orbit [19]. MMOs are also linked to slow passage through a delayed Hopf bifurcation [20] as well as to the subcritical Hopf-homoclinic bifurcation [13, 14]. Another explanation for the emergence of MMOs, stays on the so-called canard mechanism. This idea find more recently, an extention to accommodate more general classes of systems that exhibit canard dynamics [3]. It is possible to expect that the new topics will certainly emerge in a near future as MMOs seem to attract more and more mathematicians, biologists, physicians and engineers.

In the modern approach to nonlinear complex systems, not having frequently explicit solution in terms of Newton's analytical methods, the emphasis is put on qualitative dynamics rather than strictly quantitative solutions. However, deterministic MMOs are an extremely complex example of dynamic phenomena that manifest themselves as periodic, continuous non-sinusoidal oscillatory changes in the state of a system. As a prototypical example, periodic oscillations in the form of pulsations of the heart of a healthy person can be used [6]. This requires an in-depth look at how dynamic processes can develop in nature.

When the differential equations of a dynamical system are nonlinear, then their analytical solution is often unknown and they must be solved numerically with effective computer assistance. Hence, both the development of computers and the practical dissemination of numerical computations contribute significantly to the progress in research on the dynamics of all nonlinear systems. Performing computer calculations, one can get not only numerical solutions to various problems, but also discover completely unexpected phenomena. In support of this statement, an excellent example can be cited, which was the observation made in 1963 by Lorenz [21], who identified the unpredictable long-term evolution of solutions to only three nonlinear differential equations. These studies dealt with the dynamics of atmospheric phenomena. It is worth emphasizing that this discovery initiated the intensive development of research into deterministic chaos, characterized by complex, nonperiodic dynamics generated by a completely deterministic non-linear dynamical system, without the occurrence of stochastic processes.

Recall that mixed-mode dynamics is a complex type of nonlinear dynamical behavior that is characterized by a combination of small-amplitude oscillations and large-amplitude excursions of relaxation type. MMOs are frequently encountered in nonlinear systems of differential equations representing multiscale dynamical systems, in which the relevant variables evolve over several distinct scales. Consequently, typical MMO patterns in such systems consist of oscillatory waveforms in which amplitudes of different orders of magnitude alternate.

Local interactions between components of an initial disordered system create a process of self-organization leading to some form of overall order. The resulting organization is wholly decentralized, distributed over all the components of the system. The process can be spontaneous when sufficient energy is available, not needing control by any external agent. It is often triggered by seemingly random fluctuations, amplified by positive feedback.

Let us now specify four fundamental conditions of self-organization:

1. strong dynamical non-linearity, often necessarily involving ***positive*** and negative feedback,
2. balance of exploitation and exploration,
3. abundant interplay,
4. accessibility of energy to overpass natural tendency promoting increase of entropy, or disorder.

It should be emphasized that the differential equations meeting all these conditions are formulated in terms of appropriate dynamic variables, assumed as state variables, e.g., the intermediate agents concentrations, electrode potential, speed of robot's arm, etc. The occurrence of self-organization, i.e., of spontaneous oscillations of these variables depends now on the values of parameters that are included in the state variables equations creating equations in the state space of a given system. These parameters values we shall call control parameters (this may be, for instance, the flow rate through the reactor for the homogeneous chemical reactions, and the applied electrode potential or the imposed electric current for electrochemical processes and so on). It may happen that for some control parameters which we choose initially, the solutions of those equations will be quite trivial, i.e., they will predict the existence of only a single steady state. However, for another set of those parameters the same equations may generate, e.g., the variable spontaneous oscillations in time. It is now important to find the critical value of one (or more) parameter at which such qualitative change in the system's behavior occurs. This is the bifurcation point (following the Latin word "bifurcus"—forked). This topological, essentially mathematical term, refers strictly to the theory of differential equations, as they describe the dynamics of any system. Then a typical sequence of waveforms is that even relatively uncomplicated system, described in terms of deterministic differential equations, may exhibit sudden change to a completely new, qualitatively different behavior upon smooth variations of the control parameter. At the bifurcation point the system may be particularly sensitive to the fluctuations. In fact, it is the amplification of the small fluctuations to macroscopic scale which manifests itself as the qualitatively new system's behavior. This also means that in the trivial, linear range of the system's behavior the fluctuations are damped, but in a nonlinear region, due to the feedback loops, they are quickly (exponentially) enhanced. Furthermore, at the bifurcation point several possibilities for the further system's evolution may be opened and which of them will be chosen, depends on the actual random fluctuation; in this way in the essentially deterministic system, as defined by the respective differential equations, the stochastic element appears.

All pocesses of this type are strictly related to the stability of the states. When in the initially homogeneous system, a dynamical temporal or spatiotemporal order emerges, it means that the initial state lost its stability and a new behavior acquired it. If one type of self-organized phenomena turns into another one, it also means the exchange of their stability. It is very important to distinguish between the terms "steady state" and "stable state." The term "steady state" (or stationary state) means the state, the characteristics of which do not change as a function of time. In mathematics, the steady state is termed also the fixed point or the equilibrium point, although the latter term may seem a bit controversial for the physicists, chemists, biologists or mechanicians, if one considers the nonequilibrium steady state. It is further important to note that the steady state can be stable or unstable. We shall give here the simplest interpretation of these terms, which will be developed in more detail later. The stable steady state is resistant to external or internal perturbations in this way that if they happen, they are eventually damped and the system returns to its original state. The unstable steady states will not survive any fluctuation, as the system driven out of it goes then to the closest stable state. This is the reason for which the states which are observable must also be stable, since real systems are always a subject of fluctuations. Unstable steady states are not observable, unless they are stabilized by special procedures, but their existence can be indicated in theoretical models. One may additionally test the stability of the given state by introducing external excitations of increasing amplitude. If the system returns always strictly to the initial state, one calls its stability as the asymptotic one, as all the imbalances decay to zero asymptotically in a function of time. If the system characteristics survive in this way even relatively large fluctuations, covering the entire possible range of the dynamic variables considered, it is called "globally stable." Otherwise, if sufficiently large perturbations cause the switching of the system to another, concurrent stable state, the original one is called only "locally stable." More type of stability are formulated in terms of the system perturbation by relatively small fluctuations leaving the initial state, but remaining close to its characteristics. Then the system is called stable in a Lyapunov sense meaning that perturbations applied in the vicinity of the steady state cause that the system still remains in the neighborhood of this state and accepts certain tolerance for this deviation. On the other hand if this neighborhood reduces eventually to zero, then more general Lyapunov stability turns into its special case of asymptotic stability.

Note that completely different, physically or biologically, processes can exhibit the same types of bifurcations, i.e., the same types of qualitative changes of dynamics upon variation of appropriate control parameters. In the example earlier, the bifurcation involved the loss of stability of the steady state and the birth of the stable oscillatory regime, instead. Such a change of stability of the states can be an example of the Hopf bifurcation, one of the most frequent mechanisms in which the oscillations are born from the trivial, nonoscillatory steady state.

A system with one fast and one slow state variable is presented in Fig. 1.1a. It constitutes a prototypical example in the form of a van der Pol circuit and is represented by the system of equations

$$\varepsilon \dot{x} = y + ax + bx^3 \tag{1.1a}$$

$$\dot{y} = E - x \tag{1.1b}$$

Here, a > 0 and b < 0 are real constants, parameters $0 < \varepsilon \ll 1$ and E are small. The prime sign denotes differentiation with respect to time t. This slow-fast system has only one fast and one slow variable; the variable x is fast and the variable y is slow. However, it exhibits in fact complicated dynamics that are authentically surprising. In order to refine the analysis, we assume for further considerations the cubic characteristics of the current source $i(x)$ with $a = 1$ and $b = -1/3$ and switching from the slow time scale τ to the fast time scale $t = \tau/\varepsilon$.

By setting $\varepsilon = 0$ in (1.1a), we obtain the reduced system with an algebraic equation that defines the critical manifold of (1.1) as the cubic curve

$$\mathcal{M} = \{(x, y) \in R^2 | y = 1/3x^3 - x\} \tag{1.2}$$

In Fig. 1.4b the critical manifold $\mathcal{M}$ is shown as the red cubic curve divided in three segments: $\mathcal{M}^{p-}\mathcal{M}^r$ and $\mathcal{M}^{p+}$. Their limits from the right and left hand sides are fixed by points $p^- = (-1, 2/3)$ and $p^+ = (1, -2/3)$, at maximum and minimum of the manifold, respectively. The closed curve is a singular orbit composed of two fast trajectories starting at the two fold points $p^\pm$ concatenated with segments of $\mathcal{M}$.

Thus, the trajectories of (1.1) converge during fast stages to solutions of the fast subsystem called also layer equations

$$x' = y + ax + bx^3 \tag{1.3}$$

$$y' = 0.$$

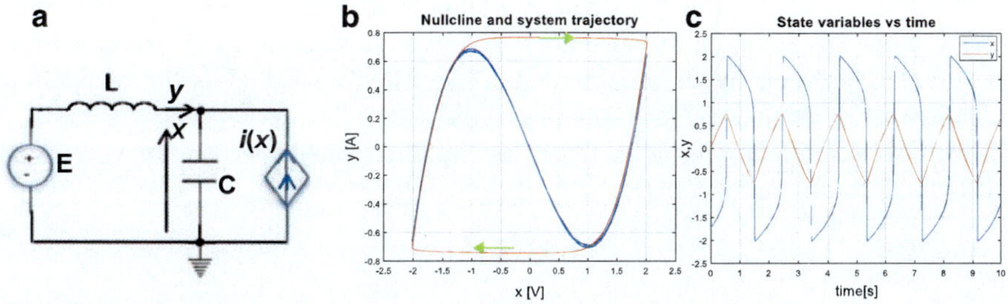

Fig. 1.4 Slow-fast system: **a** scheme of van der Pol circuit, **b** critical manifold, **c** state variables versus time

By analyzing the manifold curve in more detail, it is easy to establish that the left and right segments of $\mathcal{M}$ are attractive for oscillations in the system and the middle segment is repulsive. Moreover, $\mathcal{M}$ has a fold with respect to the fast variable x. The dynamics determined by any point external to $\mathcal{M}$ is entirely controlled by the direction of the fast variable x what is indicated in Fig. 1.4b by the horizontal double arrows.

In the case of the slow flow, on the other hand, trajectories of (1.1) converge to solutions of

$$0 = y + ax + bx^3$$

$$\dot{y} = E - x \tag{1.4}$$

which is a differential–algebraic equation (DAE) called the slow flow or reduced system.

Combining Eq. (1.4) together it is easy to detect the following relationship

$$\dot{\boldsymbol{x}} = \frac{\boldsymbol{x} - \boldsymbol{E}}{\boldsymbol{a} + 3\boldsymbol{b}\boldsymbol{x}^2} \tag{1.5}$$

The direction of the slow flow on $\mathcal{M}$ is indicated in Fig. 1.4b by the arrows on the red curve. Note, that the slow flow does depend on E, because the direction of the flow is partly determined by the location of the equilibrium at $x = E$ on $\mathcal{M}$. The slow flow is not well defined at $x = \pm 1$, but rescaling time variable with the factor $(x^2 - 1)$ leads to the desingularized reduced flow determined by the equation

$$\dot{x} = E - x \tag{1.6}$$

The singular orbit follows the slow flow on $\mathcal{M}$ to a fold point, then it jumps, that is, it makes a transition to a fast trajectory segment that flows to another branch of $\mathcal{M}$. The same mechanism returns the singular orbit to the initial branch of $\mathcal{M}$. It can be shown [1, 22, 23] that the singular orbit perturbs for $\varepsilon > 0$ to a periodic orbit of the van der Pol equation that lies $O(\varepsilon^{2/3})$ close to this candidate. Periodic orbits that alternate between stages of slow and fast oscillations called as the relaxation oscillation are presented in Fig. 1.1c.

The above considerations are the basis for explaining the generation of MMOs in a 3D layout, described by equations

$$x' = -y + ax^2 + bx^3$$

$$y' = \varepsilon(x - z), \tag{1.7}$$

$$z' = \varepsilon(E + \Psi(x, y, z))$$

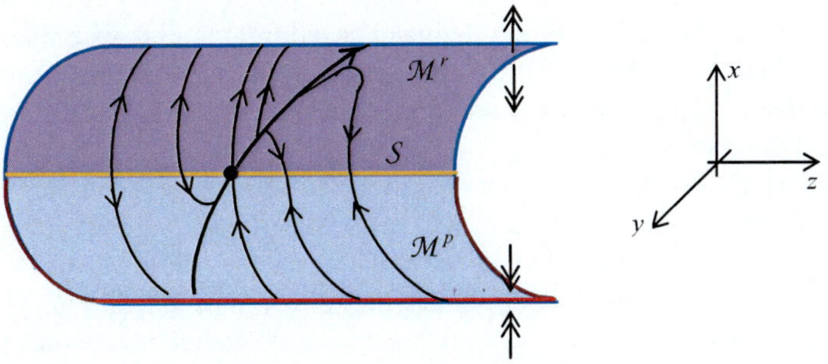

Fig. 1.5 Illustration of the slow flow on M

where ε and E are small parameters. The smooth function $\psi = 0(x, y, z)$ depends on the system state variables. Observe that in this case the new slow variable z in (1.7) takes the role of E in (1.1). Similarly as in the previous case the critical manifold M is determined by the relationship

$$y = ax^2 + bx^3 \tag{1.8}$$

Moreover, its left and right fold segments are limited by $\mathcal{M}^{p-} = \{(0,\ 0,\ z)\}$ and $\mathcal{M}^{p+} = \{(-2a/3b,\ 0,\ z)\}$ denoting the lower and upper fold lines for (1.7), respectively, and note that $p \pm$ are determined by imposing $df(x)/dx = 0$, in addition to (1.8). A repelling fold segment is given by

$$\mathcal{M}^r = \{(x, y, z) \in R^2 | 0 < x < -2a/3b,\ 0 < y < 4a^3/27b^3\} \tag{1.9}$$

The detailed structure of the MMO trajectories that is generated by (1.7) depends strongly on certain specific parameters of the system under consideration. The singular limit of $\varepsilon = 0$ in (1.7) is described by the dynamics of the reduced problem on the critical manifold $\mathcal{M}$. The properties of the global return mechanism, defined by the interplay of E and ψ in (1.7), appear as concerning aspects of the studied problem and in particular how far back the value of ψ is reset by that return.

An example of the slow flow on $\mathcal{M}$ is shown in Fig. 1.5. The thick dot on S is the folded singularity at which S changes from attracting to repelling with respect to the slow flow.

1.4 Dissection of the MMOs into Their Stages of SAOs and LAOs

Such problems as, among others, the bursting dynamics in Taylor–Couette flow, the compartmental Wilson–Callaway model of the dopaminergic neuron, stellate cell dynamics, the famous Hodgkin–Huxley model of neuron dynamics, the coupled calcium oscillator model, the complex plasma instability model, and the surface oxidation reaction and autocatalytic chemical reaction (for such examples see [5, 17]) are described by singularly perturbed systems of nonlinear ordinary differential equations (ODEs) with three (or more) state variables changing at different time scales (usually two or three).

It is worth emphasizing that when examining processes in such complex, nonlinear systems that evolve over time, it is necessary to take simultaneously into account many different factors affecting the dynamics of the whole system [1, 5, 6, 24, 25]. Their importance consists in the fact that they allow to isolate the contribution of several key parameters from the batch process, trying to predict the dominant effects, as well as changes in time of the system variables. Therefore, when examining the complexity of signals changing over time then it is valuable, without any preliminary assumptions about the nature of the phenomena analyzed, to check whether they do not contain a random component. In this way, the time taken for introductory examination of MMOs systems can be minimized, particularly at the costly pilot stage, and in result the efforts made to develop a practical investigations procedure can be significantly reduced. If such suitability can be successfully determined, then there is a chance to build a predictive model of the MMOs system.

Another way to quantify complexity of the MMOs is by entropy, which deals with the amount of information needed to predict the future state of the system. In this scope, more complex dynamics are represented as a larger entropy, and random noise exhibits maximal entropy. It is worth underlining that the entropy is a powerful information tool which can be used for identifying systems with MMOs [9, 11, 12]. It can be able to show the difference between time varying regular and irregular such as chaotic or multiple spike signals.

Since each of these categories of processes has to deal with its specific expression, then the universality of dynamic comportment is clearly manifested entirely at the level of their mathematical description, which in turn is based on the run of the solutions of the respective nonlinear ODEs. Therefore, the current achievements in the field of dynamics of nonlinear systems constitutes a significant help in the analysis of MMOs occurring in real systems.

The general form of ODEs describing dynamic states of slow-fast systems is defined as

$$F\left(z', z, u, \lambda, t\right) = \mathbf{0} \tag{1.10}$$

where $z \in R^{n}$ and $\boldsymbol{u} \in R^{m}$ are vectors of the system state-variables and forcing terms, respectively, $\boldsymbol{\lambda} \in R^{p}$ expresses system's parameters and prime sign ' denotes first derivative with respect to time $\tau \in R$. The vector function $\boldsymbol{F} \in R^{n} \times R^{n} \times R^{m} \times R^{p} \times R$ denotes a system dynamics mapping vector and contains both differential and algebraic equations. Equation (1.10) is the most general form of ODEs called a fully implicit DAE system. Note that there are also special forms of (1.10) such as fully-explicit or semi-explicit forms and special system in Hessenberg form [26]. They all, after some usually simple manipulations take just form of ordinary differential equations (ODEs) and typical mathematical software packages, such as Matlab, Maple and Mathematica can be applied as their effective solvers.

Few attempts have been made thus far to classify different forms of nonlinear ODEs, in contrast to the classification of the related phenomena of linear system oscillations. One of the research currently under development in this field is nonlinear systems showing combinations of state variables with fast and slow variation in time.

The dynamic states of such systems are described by a slow-fast vector field that takes the form of singularly perturbed systems of ODEs which are closely related to semi-explicit DAEs, namely

$$\varepsilon \dot{\boldsymbol{x}} = \boldsymbol{f}(\boldsymbol{x}, \boldsymbol{y}, \lambda, \varepsilon) \tag{1.11}$$

$$\boldsymbol{y} = \boldsymbol{g}(\boldsymbol{x}, \boldsymbol{y}, \lambda, \varepsilon)$$

where $\boldsymbol{x} \in R^m$ and $\boldsymbol{y} \in R^n$ are two vector components of the system state-variables vector, $\lambda \in R^p$ are circuit parameters, and ε is a small parameter being in the range $0 < \varepsilon \ll 1$ which represents the ratio of time scales and it makes the system singularly perturbed. The vector functions $\boldsymbol{f}: R^m \times R^n \times R^p \times R \to R^m$ and $\boldsymbol{g}: R^m \times R^n \times R^p \times R \to R^n$ are assumed to be sufficiently smooth, typically C^∞. The over dot stands for the natural, i.e., slow time scale τ. The components of the vector $\boldsymbol{x}$ are fast and that of the vector $\boldsymbol{y}$ are slow.

By switching from the slow time scale τ to the fast time scale $t = \tau/\varepsilon$ Eqs. (2.1) can be transformed to

$$\boldsymbol{x}' = \boldsymbol{f}(\boldsymbol{x}, \boldsymbol{y}, \lambda, \varepsilon) \tag{1.12}$$

$$\boldsymbol{y}' = \varepsilon \boldsymbol{g}(\boldsymbol{x}, \boldsymbol{y}, \lambda, \varepsilon)$$

where prime sign ' denotes first derivative with respect to time $t \in R$. Solutions of Eqs. (1.12) frequently exhibit slow and fast stages characterized by the different speed at which the solution advances. As $\varepsilon \to 0$, the trajectories of (1.12) converge during fast stages to solutions of the fast subsystem or layer equations.

The procedure schematically presented in Fig. 1.6 seems appropriate in the process of identifying the conditions for the formation of MMOs.

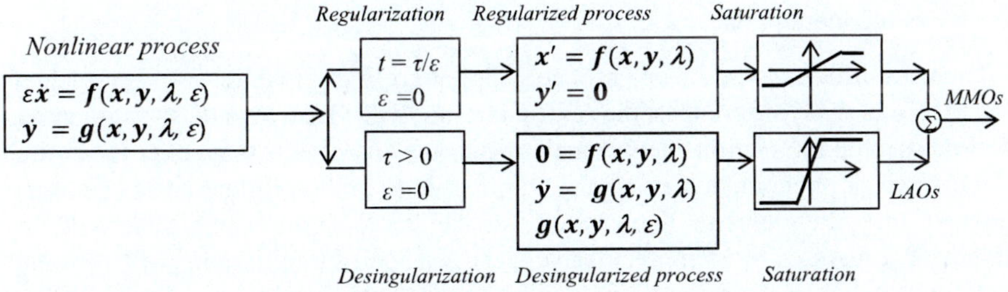

Fig. 1.6 Scheme of the dissection procedure

The key feature in studying this type of procedure is that that the dynamics of $\boldsymbol{x}$ and $\boldsymbol{y}$ can be considered separately [13]. It should be easily highlighted that the rate of change of $\boldsymbol{y}$ defined by (1.12) does not depend on $\boldsymbol{x}$. However, it seems justifiable to suppose that certain bursting oscillations in alive organisms can be forced by external periodic excitations that are independent of this process, as in the case of a circadian cycle, for example.

1.5 Emergence of Mixed Mode Oscillations

The popularization of large-scale systems and inexpensive robust installations leads to the need to treat them as extensive systems. For this reason, it is advisable to focus on the operation of individual devices in the system or even entire subsystems. This leads to the complex object being treated as a modular system that can be easily configured and maintained in proper operation compared to a monolithic system. If a combination of a large number of simple elements as one object is presented, then its operation is more complex than the operation of only the elements. The functioning of the whole system results not only from the functioning of each of its parts separately, but also from the mutual relations and interactions of these parts with each other. Thanks to this, it is possible to build systems with a much richer behavior and functioning than it would only result from their components. This coming to light of qualitatively new forms and behaviors from the interaction between simpler elements is called emergence. Generally, this term is explained by saying that the effect of the action of the whole is not only the sum of the effects of the action of its parts—as a whole, there are usually new properties and phenomena that do not appear in its components. A system is defined as emergent if in some sense it cannot be represented by describing only its component parts. This means, in the strict scientific sense, that the given object is irreducible. In this case, the simplified description of the lower level object known to us is not sufficient to describe the higher level object. In others words an emergent system is one where:

1. the final system has properties and effects which exceed the properties and effects of its component parts, and
2. whose final properties and effects cannot be predicted from merely examining the component parts.

There are different types of emergence and related effects. In the case of emergence related to size scale differences, the reason is usually the ordering and strengthening of some specific interactions between elements, causing effects that cover the entire system. Such a phenomenon is usually impossible to predict on the basis of observation of only the elements themselves. Air can serve as an example, as none of the physical features of its molecules suggest that their large number will form an object (air) capable of transmitting sound waves. Another example is the so-called the "wetness" of water (H_2O). It results from the fact that water has a certain set of properties, which we commonly refer to when we say that water is "wet", although its components—hydrogen and oxygen—are not. Well, this “wetness” of water comes from the fact that the hydrogen and oxygen atoms in the water molecule interact so that an elongated structure is formed, one end of which has a positive electric charge and the other negative one. Hence the water molecule is an electric dipole. Thanks to these electrically charged extremities, water molecules can interact in a special way with each other and with molecules of other substances, in a way that we attribute to the “wet” quality. However, at least 16 molecules H_2O must combine into a single system for this “wet” property to occur. In the case of separate molecules of hydrogen (H_2) and oxygen (O_2), their charges are evenly distributed, that is, they are not dipoles and therefore are “dry". One of the reasons why the emergent behavior of systems is difficult to predict is the number of interactions between elements that should be considered when simulating them. Often it is sufficient to consider only very simple interactions: for example, simply identifying the reaction of each locust insect to the behavior of its neighbors is sufficient to simulate the movement of an entire locust colony. The same is not possible with air, however, due to the astronomical number of molecules to be simulated. However, a multiplicity of interactions alone is not enough to produce the emergent behavior of a system. If they are not properly related, they can mutually cancel out, generating only noise on a larger scale. Systems with emergent properties can appear to be non-compliant with the second law of thermodynamics because they self-order without a central source of control. In reality, however, this principle is not broken by them, because their operation always takes place at the expense of releasing entropy to the environment. In a frictional system individual atoms do not create friction, because all the forces they are subject to are conservative. Only the proper arrangement of atoms into a crystal structure enables the conversion of mechanical energy into heat, which we define as friction. Similar descriptions apply to other emergent systems that exhibit properties such as viscosity, elasticity, strength, etc. In classical mechanics, the laws of systems can be described as emergent, and resulting from averaging the principles of quantum mechanics over all elements of a macroscopic object. This is of particular importance as quantum mechanics is usually considered more complex, while in most cases we assume that lower-level rules are less complex than the resulting emergent properties.

A similar dependence occurs in the case of the laws of thermodynamics. They are relatively simple, although they result from complex interactions between elementary molecules of the system. It should be emphasized that the emergent behavior of the system is the result of non-linear interactions of elements causing deterministic processes. Such behavior should be distinguished from accidental, as their nature is different. Various forms of nonlinearity are the cause of emergent states of practical systems. Three of the possible ones are the most important, namely: technological limitations in the production of high-capacity components, feedback and signal transmission delays. The emergent system can operate in a wide range of dynamic states, including the undesirable steady states, oscillations and chaos. Therefore, the first step in analyzing such cases of systems is to identify whether any of these particular states are present in the system under study. Let the process of supplying parts on the assembly line in a certain car factory serve as an illustration of this issue. We assume

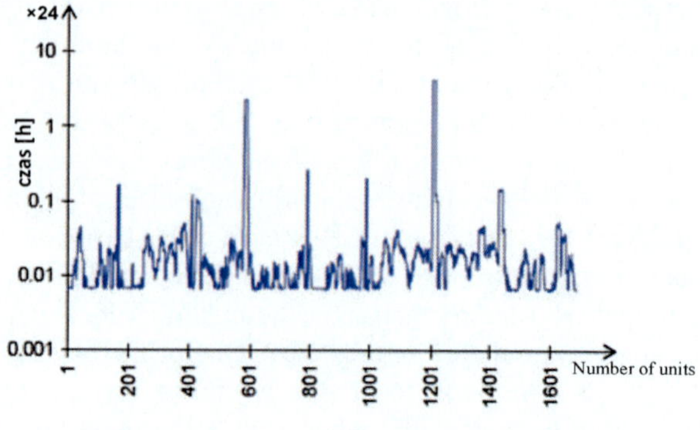

Fig. 1.7 Dependence of the delay time on the number of units produced

that the assembly takes place on the production line and the appropriate sensors are installed at points that are critical for assembly. The result of their operation is the measurement of the assembly delay time caused by the delivery of parts. More than 1600 units coming off the belt over a 14-day period were analyzed. The measured delay times are shown in Fig. 1.7. The axis of ordinates shows the delay times on a logarithmic scale, depending on the production progress.

It can be clearly seen changes in lag times on a scale by three orders of magnitude. Delay times greater than 2 h occur almost regularly, but shorter delays are not regular. Such dependencies make it possible to identify potential cases of emergent production systems. If emergent operation of a given system is not desired, then an appropriate strategy of its control should be adopted. One of them is to eliminate nonlinear interactions from the system, which is often not feasible. Another way is to cause any oscillation in the system to be dampened by limiting its operation well below the optimality level. In the case of automation, emergent states of systems appear to be a consequence of modern architecture of control systems [27, 28]. Designers and producers of appropriate technologies must take certain steps to make it easy to identify, evaluate and modify the resulting systems. This provides opportunities to reduce potential damage due to unpredictable system behavior. Consequently, a system is obtained which is more efficient and resistant to undesirable states.

Diverse neuronal oscillators arising in a mixed population of neurons with different excitability properties produce mixed mode oscillations (MMOs). Various MMOs, including MMBOs (mixed mode bursting oscillations) and synchronized tonic spiking appear in a randomly connected network of neurons, where a fraction of them is in a quiescent (silent) state and the rest in self-oscillatory (firing) states [8]. MMOs and other patterns of neural activity depend on the number of oscillatory neighbors of quiescent nodes and on electrical coupling strengths. For weak neuron couplings, MMOs appear due to the desynchronization of a large number of quiescent neurons in the networks. The quiescent neurons together with the firing neurons generate oscillations of high frequency and bursting mobility. The overriding goal is to detect a favorable network architecture of neurons and corresponding parameter spaces where Izhikevich model [3] of neurons produce diverse responses laying from MMOs to tonic spiking. In the absence of neuron coupling, the activity of the considered neuronal population reveals two types of dynamical states (or excitabilities), ranging from spike-bursting by subthreshold to quiescent states. The coupling parameter space and the ratio of mixed populations where MMOs and fast tonic spiking exist create basic factor to form emergence. At weak couplings of neurons and a diluted random network setting, the desynchronized subthreshold neurons exhibit MMOs. With the increase of the coupling, all subthreshold neurons fire in a mixed-mode state. In both cases, MMOs are not prominent in oscillatory neurons and eventually disappear as the coupling strength increases. Consequently, neural subpopulations emerge as synchronous clusters exhibiting tonic spiking behavior. For diluted random and homogeneous networks, where the electrical coupling strength is constant, the neighbors exhibiting self-sustained oscillations determine the structural patterns of MMOs. A given mixed population of quiescent and oscillatory nodes can provide rise to several types of MMOs and MMBOs in the two types of neuronal networks. MMOs take potential applications in biophysical, medical and other systems. In complex systems, various mechanisms subsist during different oscillatory phases generating spike patterns between fast and slow amplitude motion together with spikes and subthreshold oscillations. There exists a suggestion that MMOs can be responsible for the transition from high firing rates to quiescent states by reducing neuronal gain and synaptic plasticity. It is also noticed [8] the impacts of small amplitude oscillations/subthreshold oscillations (STOs) on diverse neuronal responses such as spike clustering, rhythmic activities, synchronization and others as emergent processes.

References

1. Desroches, M., Guckenheimer, J., Krauskopf, B., Kuehn, C., Osinga, H.M., Wechselberger, M.: Mixed-mode oscillations with multiple time scales. SIAM Rev. **54**, 211–288 (2012)
2. Fechner, G.T.: Massbestimmungen über die galvanische Kette. Zeitschrift fur Chem. und Phys. **53**, 141 (1828)
3. Brunel, N., Hakim, V.: Sparsely synchronized neuronal oscillations. Chaos **18**, 015113 (2008)
4. Erchova, I., McGonigle, D.J.: Rythms of the brain: an examination of mixed mode oscillation approaches to the analysis of neurophysiological data. Chaos **18**, 015115 (2008)
5. Focus issue: 'Mixed-mode oscillations: experiment, computation, and analysis. Chaos **18**, 015101
6. Trzaska, Z.: Study of mixed-mode oscillations in a nonlinear cardiovascular system. Nonlinear Dyn. **100**(3), 2635–2656 (2020)
7. Marszalek, W., Trzaska, Z.: Mixed-mode oscillations in a modified Chua's circuit. Circ. Syst. Sig. Process. **29**(6), 1075–1087 (2010)
8. Ghosh, S., Mondal, A., Ji, P., Mishra, A., Dana, S.K., Antonopoulos, Ch.G., Hens, Ch.: Emergence of Mixed Mode Oscillations in Random Networks of Diverse Excitable Neurons: The Role of Neighbors and Electrical Coupling. Frontiers in Computational Neuroscience, Vol. 14, Article 49 (2020)
9. Rotstein, H.G.: Mixed-mode oscillations in single neurons. J. Dyn. Diff. Equat. **27**, 83–136 (2015)
10. Blagojević, S.N., Čupić, Ž., Ivanović-Šašić, A., Kolar-Anić, Lj.: Mixed-mode oscillations and chaos in return maps of an oscillatory chemical reaction. Russ. J. Phys. Chem. A **89**(13), 2349–2358 (2015)

11. Brons, M., Kaasen, R.: Canards and mixed-mode oscillations in a forest pest model. Theoret. Popul. Biol. **77**, 238–242 (2010)
12. Marszalek, W., Trzaska, Z.: Mixed-mode oscillations and chaotic solutions of *jerk* (Newtonian) equations. J. Comput. Appl. Math. **262**, 373–383 (2014)
13. Berglund, N., Gentz, B., Kuehn, Ch.: From random poincaré maps to stochastic mixed-mode-oscillation patterns. J. Dyn. Diff. Equat. **27**, 83–136 (2015)
14. Brøns, M., Krupa, M., Wechselberger, M.: Mixed mode oscillations due to the generalized canard phenomenon. Fields Inst. Commun. **49**, 39–63 (2006)
15. Marszalek, W., Trzaska, Z: Memristive circuits with steady-state mixed-mode oscillations. Electron. Lett. **50**(18), 1275–1277 (2014)
16. Doedel, E.J., Pando, C.L.: Rare events in mixed-mode oscillations from weakly coupled lasers. Phys. Rev. E **100**, 052204 (2019)
17. Ostwald, W.: Periodic phenomena in the disintegration of chrome in acids. Z. Phys. Chem., Stoechiom. Verwandtschaftsl. **35**, 33–76 (1900)
18. Krupa, M., Popovic, N., Kopell, N.: Mixed-mode oscillations in three time-scale systems: a prototypical example. SIAM J. Appl. Dyn. Syst. **7**, 361–420 (2008)
19. Krupa, M., Szmolyan, P.: Relaxation oscillation and canard explosion. J. Differ. Eqs. **174**, 312–368 (2001)
20. Marszalek, W.: Circuits with oscillatory hierarchical Farey sequences and fractal properties. Circ. Syst. Sig. Process. **31**, 1279–1296 (2012)
21. Lorenz, E.: Deterministic nonperiodic flow. J. Atmos. Sci. **20**, 130–141 (1963)
22. Chua, L.O., Lin, P.-M.: Computer Aided Analysis of Electronic Circuits, Algorithms and Computational Techniques. Prentice-Hall Englewood Cliffs (1975)
23. Kuehn, C.: PDE Dynamics: An Introduction, SIAM, in the series: Mathematical Modeling and Computation (2019)
24. Berglund, N., Gentz, B., Kuehn, C.: From random Poincare maps to stochastic mixed-mode-oscillation patterns. J. Dyn. Diff. Equat. **27**(1), 83–136 (2015)
25. Iuorio, A., Kuehn, C., Szmolyan, P.: Geometry and numerical continuation of multiscale orbits in a nonconvex variational problem. Discr. Contin. Dyn. Syst. S **13**(2) (2020)
26. Friedberg, S.H., Insel, A.J., Spence, L.E.: Linear Algebra. Prentice Hall, Englewood Cliffs (1989)
27. Podhaisky, H., Marszalek, W.: Bifurcations and synchronization of singularly perturbed oscillators: an application case study. Nonlinear Dynam. **69**, 949–959 (2012)
28. Jardon Kojakhmetov, H., Kuehn, C.: A survey on the blow-up method for fast-slow systems. Contemporary Mathematics, AMS, arXiv:1901.01402

Chapter 2
Analysis of MMOs in Electric Circuits and Systems

2.1 Models of Nonlinear Dynamic Circuits

In general, a nonlinear active circuit has the multipoles structure shown in Fig. 2.1. Dynamic multi-terminal capacitive $\boldsymbol{C}$ and inductive $\boldsymbol{L}$ elements (possibly nonlinear) are connected with multiport structure N composed of independent and controlled sources $\boldsymbol{e}, \boldsymbol{j}_s$ and elements described by some nondynamic and nonlinear i-v (current–voltage) characteristics (resistors, diodes, memristors, transistors, etc.). Dynamic elements consisting of C_r capacitors, $r = 1, 2, \ldots, k$, and L_m coils, $m = k + 1, k + 2, \ldots, n$ can be connected with each other and with independent and controlled sources e_s, $s = 1, 2, \ldots, p$, j_h, $h = 1, 2, \ldots, q$ as well as with nondynamic and nonlinear i-v characteristics components.

Voltages across the capacitors and currents in coils form the state variables vector denoted by $\boldsymbol{x}(t) = [v_1(\mathrm{t}), v_2(\mathrm{t}), \ldots, v_k(t), i_{k+1}(t), i_{k+2}(t), \ldots, i_n(t)]^T$, where T stands for transposition, and its components fulfill Kirchhoff voltage and current laws, and relationships determining all elements, respectively, for the whole circuit. Moreover, taking into account i-v (current–voltage) characteristics of the circuit elements one can describe the instantaneous state of the circuit by the following (written in matrix form) implicit system of equations

$$F\left(x', x, u, t\right) = \mathbf{0} \tag{2.1}$$

where $\boldsymbol{x} \in R^n$ is a state variable vector, $\boldsymbol{u} \in R^m$ is a vector expressing forcing terms (input signals), $\boldsymbol{F} \in R^p \times R^m \times R$ denotess a circuit mapping vector and prime sign $'$ denotes first derivative with respect to time $t \in R$.

Equation (2.1) are in the implicit state form and that represents the instantaneous state of the circuit. Implicit equations have been used in practice for a few decades in computer-aided modeling of large-scale systems and circuits. This is due because conversion to explicit form generally increases total parameter count. Moreover, the required implicit or explicit matrix inversions may be ill-conditioned, and therefore the conversion is simply unnecessary for many purposes [21–26, 29, 30].

Z. Trzaska, *Mixed Mode Oscillations (MMOs)*, Studies in Systems, Decision and Control 374, https://doi.org/10.1007/978-3-030-76867-6_2

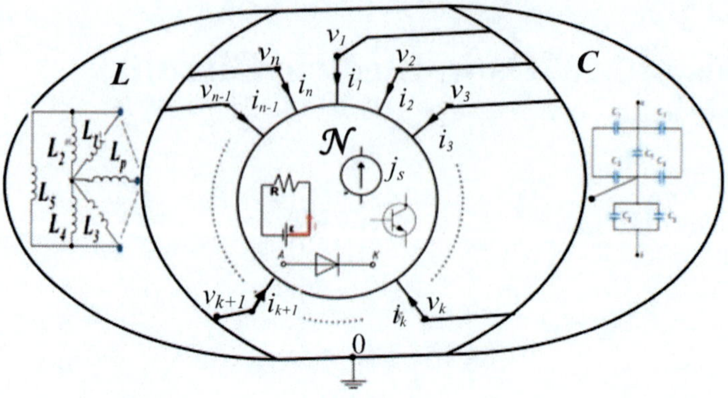

Fig. 2.1 Schematic representation of nonlinear dynamical circuits

There are three basic reasons for writing the equations in this form: (1) this form lends itself most easily to analog and/or digital computer programming, (2) the analysis of nonlinear and/or time-varying circuits is quite easy (whereas an extension is not easy in the case of loop, mesh, cut-set, or node analysis), and (3) in this form a number of system theoretic concepts are readily applicable to circuits [1].

It is worth to mention that all dynamical problems in the real world are initially modeled as (2.1). Then, most of such equations, after additional manipulations may be transformed, although frequently with hard effort, into just ordinary differential equations (ODEs). Then they are usually solved by numerical integrators designed exclusively for ODEs. However, there is a large class of problems for which such transformation is not possible. Those problems must be treated as differential algebraic equations (in brief DAEs). Additionally, even if the original problem does not seem to be modeled by a DAE, it may eventually end up as a DAE problem. The requirement comes down to the fact that (2.1) can be solved in the sense that it has a family of solutions that are uniquely defined by the parameters values of the circuit elements and of the state variables at a given $t = t_0$. These solutions have to form a manifold of integral curves, shortly the solution manifold. Singularly perturbed systems of ODEs are closely related to semi-explicit DAEs and used in the analysis of MMOs.

2.2 Basic Mechanisms of MMOs Generation in Electronic Circuits

Electronic circuits exhibiting various types of MMOs are characterized by a presence of a small parameter that makes the circuits singularly perturbed. A cubic nonlinearity is typical in circuits with MMOs. In more complicated cases the singularly perturbed systems of nonlinear ODEs may have three (or more) variables changing at three (or more) different time scales. In the recent years, the most accepted and popular approach to explain the MMOs phenomenon in R^3 is that the MMOs result from a combination of canard solutions around a fold singularity and relaxation spikes coupled together by a special global return mechanism [2–4]. Most of the solutions of MMOs are analyzed in the subsequent chapters in more detail. However, the topics and approaches presented there do not exhaust all possible discussions about MMOs and their areas of application [33]. The new questions will certainly emerge in a near future as MMOs seem to attract more and more mathematicians and specialists working in various applied sciences.

The generation of MMO time series requires the coordinated action of various mechanisms: (*i*) a mechanism for the generation of the subthreshold regime (STOs); (*ii*) a spiking mechanism, including the onset of spikes and the description of the spiking dynamics; and (*iii*) a return mechanism from the spiking regime back to the STOs regime.

A typical sequence of events is that even relatively uncomplicated system, described in terms of deterministic evolution Eq. (2.1), may exhibit sudden change to a completely new, qualitatively different behavior upon smooth variations of the control parameter. However, a too simple deterministic approach, assuming the smooth response of the system's behavior upon increasing distance from equilibrium, must be replaced here by a far more sophisticated statement, accepting the existence of sudden changes, mathematically called the bifurcations (following the Latin word "bifurcus"—forked). This, essentially mathematical term taken from topology, refers strictly to the theory of differential equations, as they describe the dynamics of any system. At the bifurcation state the circuit may be particularly sensitive to the irregular rising and falling in value of time series. In fact, it is the amplification of the small value to macroscopic scale which manifests itself as the qualitatively new circuit's behavior. This also means that in the simple, linear range of the circuit's behavior the risings are damped, but in a nonlinear case, due to the feedback loops, they are quickly (frequently exponentially) enhanced. Moreover, at the bifurcation point several possibilities for the further circuit's evolution may be opened and which of them will be chosen, depends on the actual random rising. In this way in the essentially deterministic circuit, as defined by the respective differential equations, the stochastic element may appear.

This is more than just a simple statement because this is a new aspect of the mode in which dynamic processes may develop in nature. In particular, as the source of dynamic instabilities should be considered the characteristics of not only the individual element, but of the entire electric circuit.

2.3 Mixed-Mode Oscillations in Chua's Circuits

Singularly perturbed systems of ODEs can be used in the analysis of mixed-mode oscillations produced by the so-called modified Chua's circuits made up of the coupled R, L, C elements connected with controlled sources $(1 + \gamma)y$-c and nonlinear

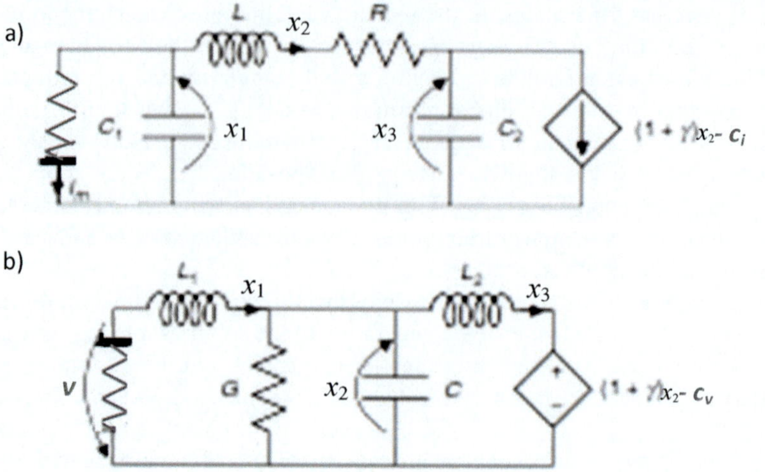

Fig. 2.2 Schemes of dual modified Chua's R, L, C circuits with: **a** lossy coil R, L, **b** lossy capacitor G, C

resistors with voltage-current characteristic $i = a_i x^2 + b_i x^3$ for circuit shown in Fig. 2.2a (equivalently $v = a_v x^2 + b_v x^3$ for the dual circuit), respectively. A resistor with cubic nonlinearity is typical in circuits with MMOs. It is worth noticing that recently, the most accepted and popular approach to explain the MMOs phenomenon in R^3 is that that they result from a joining of canard solutions around a fold singularity and relaxation spikes coupled together by a special global return mechanism [5]. The following ODEs describe each of the circuit of Fig. 2.2

$$
\begin{aligned}
\varepsilon x_1' &= -x_2 - (ax_1 + bx_1^2)x_1, \\
x_2' &= \alpha(x_1 - Kx_2 - x_3), \\
x_3' &= c - \beta x_2
\end{aligned}
\tag{2.2}
$$

where the symbol ' means differentiation with respect to time t ∈ R, $0 < C_1 \equiv \varepsilon \ll 1$, $\alpha = L^{-1}$, $K = R$ and $\beta = \gamma/C_2$ in the case of the circuit from Fig. 2.2a and $0 < L_1 \equiv \varepsilon$ $< < 1$ and $\alpha = 1/C$, $K = G$ and $\beta = \gamma/L_2$ for the circuit in Fig. 2.2b. The parameters of the controlled sources in both cases of the circuit depend on the variable y according to the expression $(1 + \gamma)$y-c with $\gamma > 0$ and c = const. The $a = a_i$, $b = b_i$ and $c = c_i$ for the first circuit and $a = a_v$, $b = b_v$ and $c = c_v$ for the dual one.

Circuit described by (2.2) is a prototypical example for MMOs examined in [5]. Its behavior depends on all five constant parameters engaged. One can distinguish three different modes of oscillations exhibited by (2.2): only small amplitude oscillations (SAOs) within the time, large amplitude oscillations (LAOs), and a composition of both SAOs and LAOs producing the MMOs phenomenon. The three modes are illustrated in Fig. 2.3 which shows the solutions of (2.2) for $K = 0$ and $\varepsilon = 0.01$, $a = 1.5$, $b = -1$, $\beta = 0.005$ and three values of c. In the SAOs case only, the small

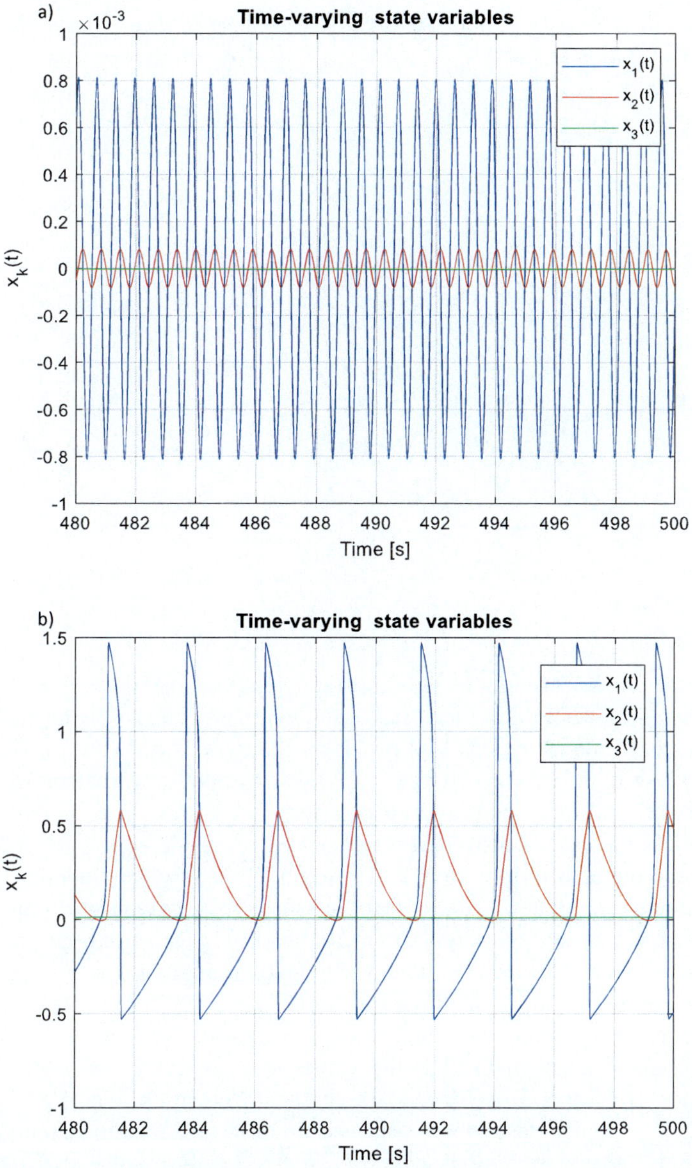

Fig. 2.3 Periodic oscillations produced by (2.2) for $\varepsilon = 0.01$, $a = 1.5$, $b = -1$, $\beta = 0.005$ and different values of parameter c: **a** SAOs only for $c_1 = 0$, **b** LAOs only for $c_2 = 0.00105$, **c** MMOs for $c_3 = 0.00053$

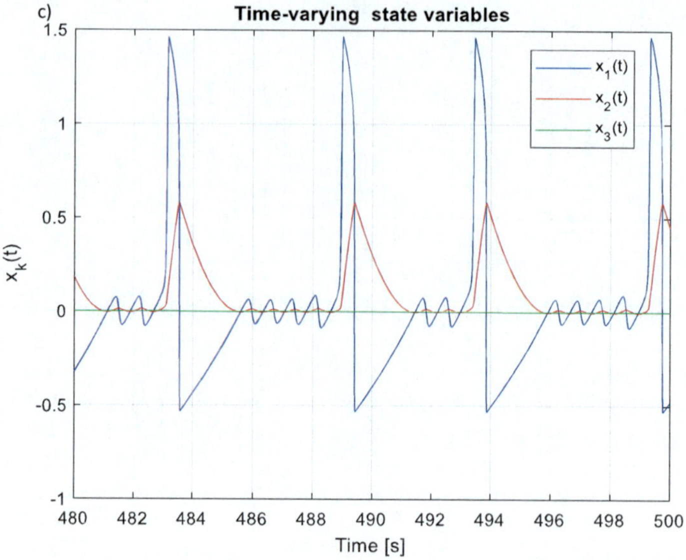

Fig. 2.3 (continued)

amplitude oscillations around the equilibrium point at the origin (0, 0, 0) are due to Hopf bifurcation for $c = 0$. In the LAOs case only, a trajectory passing close to the origin bypasses the region of small amplitude oscillations. The MMOs case is in some sense a synthesis of the previous two cases. It is worth underlining that the mechanism in which small and large amplitude oscillations occur in a same circuit is quite complex [5]. In the case of MMOs, a series of SAOs considered as canard solutions appear around the equilibrium point at the origin (0, 0, 0) which undergoes a rapid canard explosion yielding an LAOs. But they through a special return mechanism bring back the circuit into the neighborhood of the origin. Depending on the circuit's parameters, the oscillation may continue with one (or more) LAOs, or may go through a new series of SAOs after which the trajectory leaves again the neighborhood of the origin and the phenomenon reduplicates. Illustrations are presented in Figs. 2.4 and 2.5. For fixed a and b in (2.2) the values of c, β and ε are responsible for various types of MMOs. The MMOs are characterized by the Farey sequence … $L_{k-1}^{s_{k-1}} L_k^{s_k} L_{k+1}^{s_{k+1}}$ …, which describes the succession of large relaxation excursions and small (nonrelaxation) oscillations where L_i and s_i are the numbers of LAOs and followed SAOs, respectively. Figure 2.6 illustrates many typical cases that one may obtain from (2.2).

The diagrams were obtained for the fixed values of $b = 0.005$, $\varepsilon = 0.01$, $\alpha = 1.5$ and $\beta = -1$ and varying parameter c. Note that due to the fact that $a \sim O(\varepsilon^{-1})$, $b \sim O(\varepsilon^{-1})$, the $x_3(t)$ is the slowest of all three variables and $x_1(t)$ is the fastest one. With the above chosen parameters, circuit (2.2) may be called a three time-scale circuit.

The complete analysis of the occurrence of MMOs, particularly with regard to the Farey sequence L^s for (2.2), is not finished at the present time. However, several

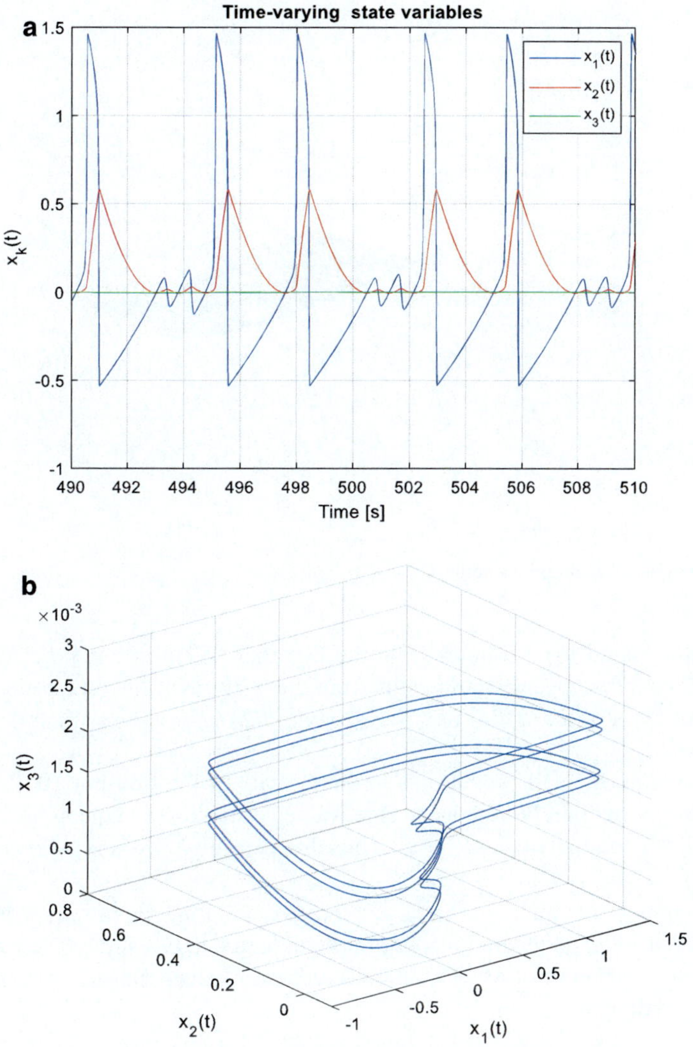

Fig. 2.4 MMOs $2^2 2^2$ produced by (2.2) for the same parameters as in Fig. 3.3 but with $c = 0.00075$: **a** time series of state variables, **b** 3D trajectory

known facts incorporating the latest developments and trends in evaluations of MMOs may be helpful in the profound analysis of (2.2). These facts are as follows [6]:

(i) MMOs occur in (2.2) if the equilibrium point at origin (0, 0, 0) is a folded node of the corresponding desingularized equations. This happens if

$$4\alpha^3 b/(27b^2) - 1/(8a) < c < a^3 b/(18b^2). \tag{2.3}$$

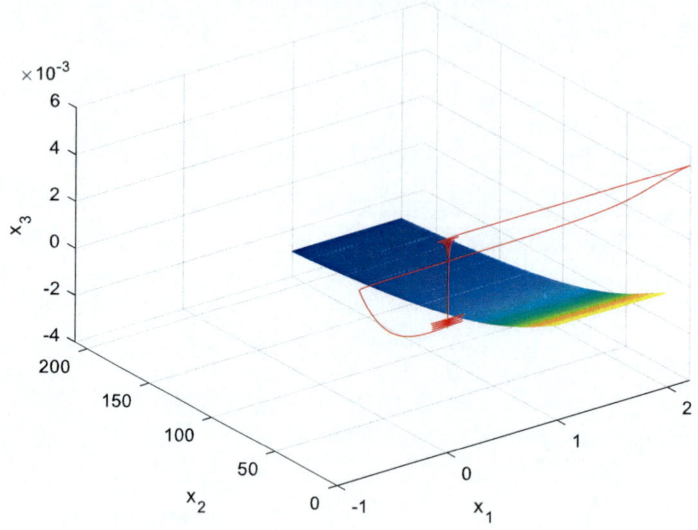

Fig. 2.5 3D plot of the circuit trajectory

The lower bound for c follows from the fact that (2.2) has a node at (0, 0, 0), see Fig. 2.7, and the upper bound results from the analysis of the return mechanism described in [5]. For $c > a^3\beta/(18b^2)$ the system (2.2) is in a pure relaxation mode with LAOs only.

(ii) the period of small oscillations in the vicinity of the fold line $(0, 0, x_3)$ can be estimated by the purely imaginary eigenvalues of linearized equations for (2.1), from which we obtain $\omega = \sqrt{\varepsilon^{-1} - \beta}$. Thus, the period of SAOs is estimated to be $T = 2\pi / \sqrt{\varepsilon^{-1} - \beta}$.

(iii) the number of small amplitude oscillations, and the type of sequence L^s, for fixed α and β depend on the initial conditions and parameters a and b. The numerical results of the occurrence of MMOs and the average s values for (2.2) are presented in the next section.

2.4 Properties of MMOs in Chua's Circuits

The detailed structure of the MMO trajectories that will be observed in (2.2) depends strongly on certain features of the specific parameter values under consideration. One important aspect concerns the properties of the global return mechanism, defined by the interplay of c and β in (2.2), and in particular how far back the value of x_3 is reset by that return. To derive asymptotic formulae for the return of trajectories under the flow of (2.2), we will define the corresponding return map on suitable sections for the flow, which we introduce below.

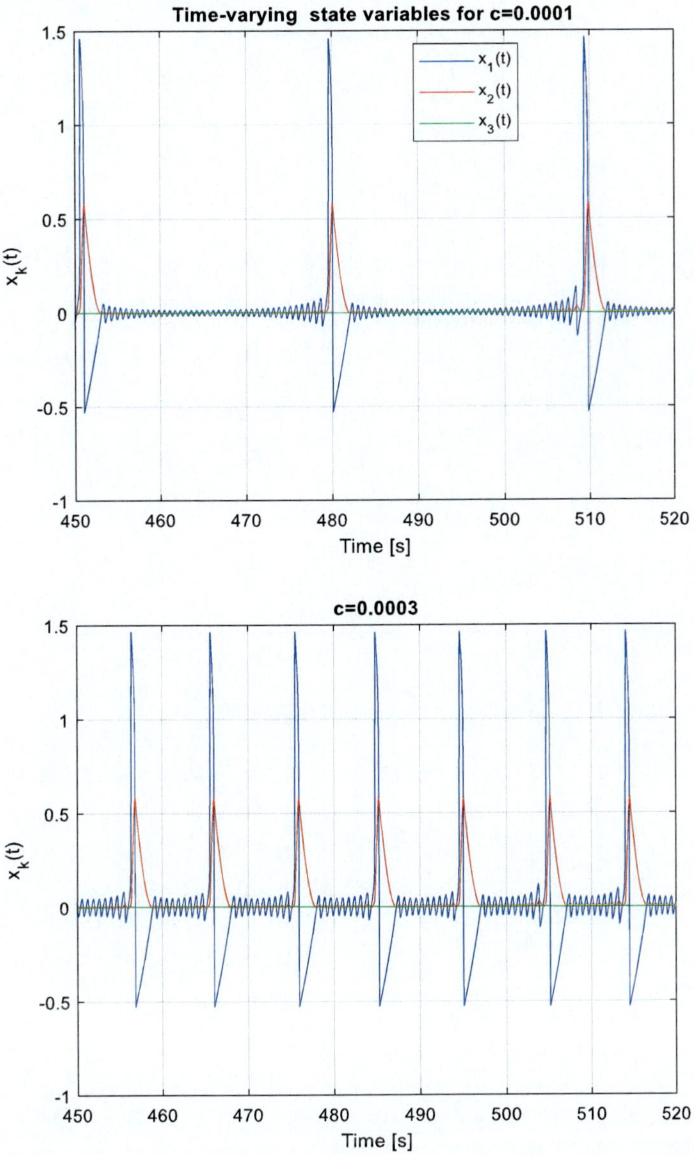

Fig. 2.6 Oscillations produced by (2.2) for various values of parameter c (other circuit parameters are same as in Fig. 2.3)

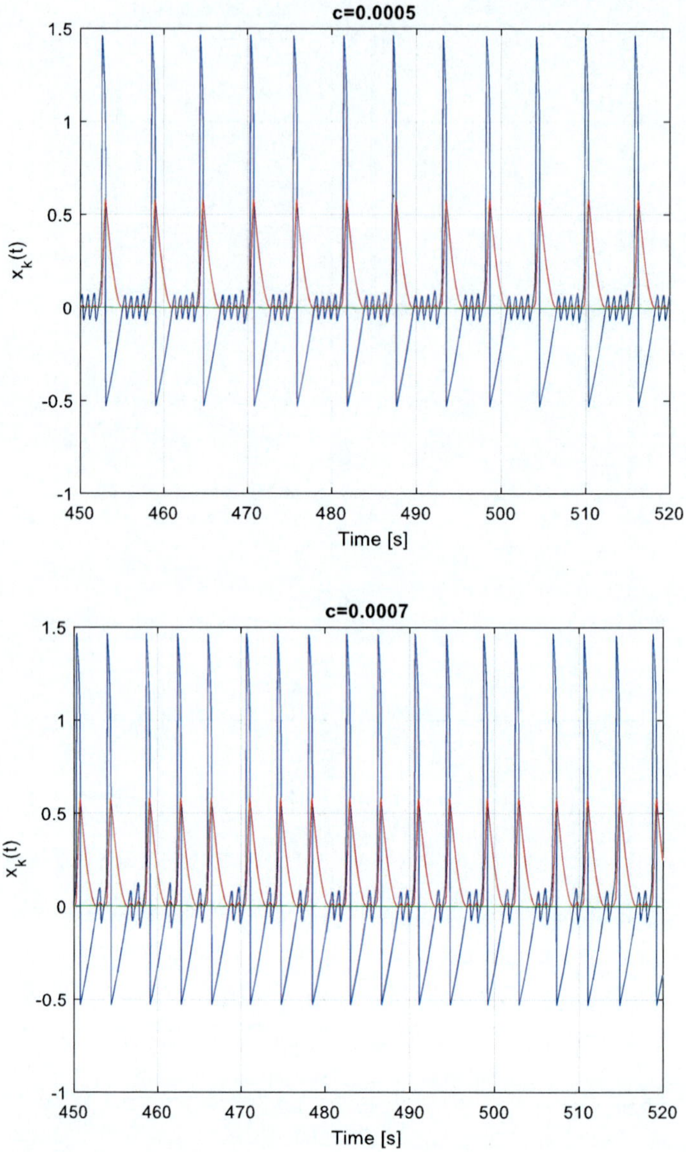

Fig. 2.6 (continued)

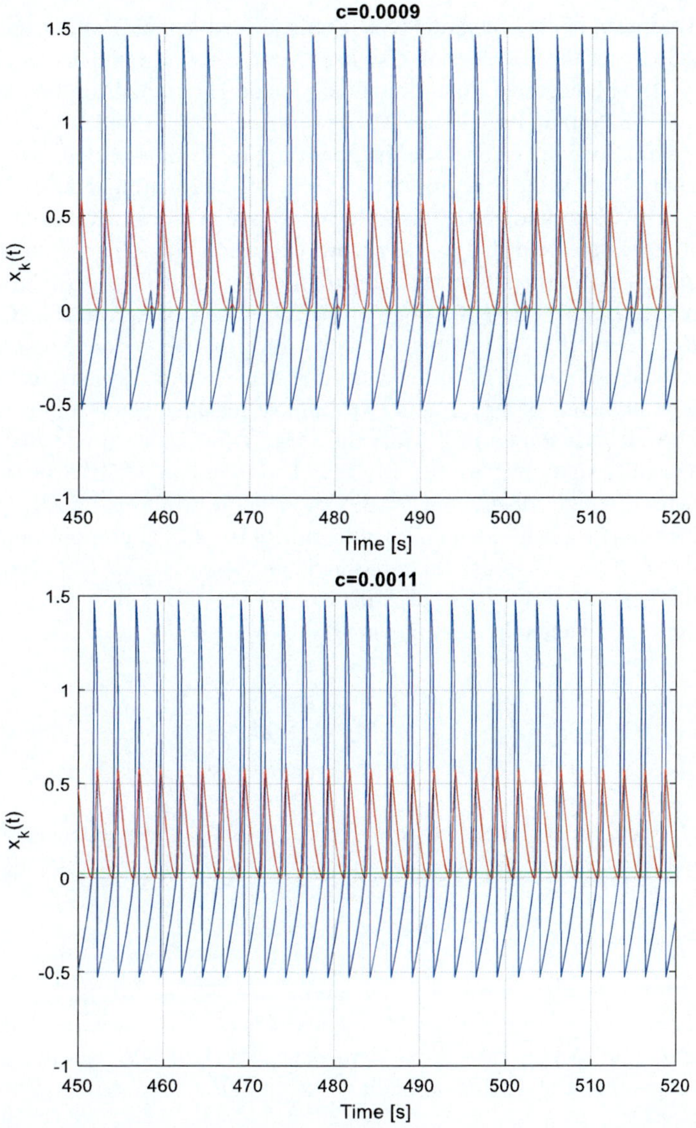

Fig. 2.6 (continued)

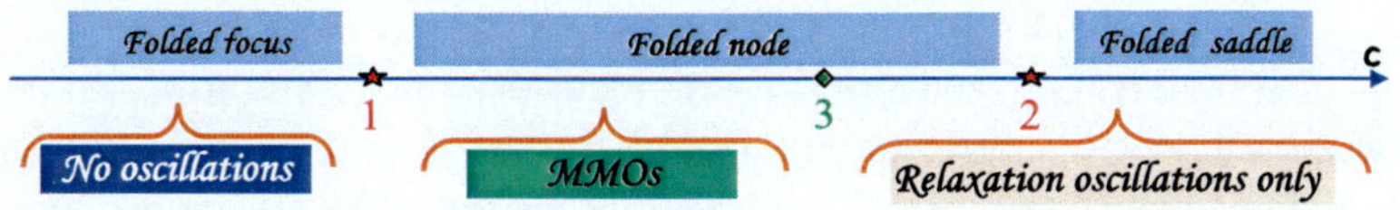

Fig. 2.7 Spans of c for various natures of responses of (2.2) with $1 = 4\alpha^3\beta/(27b^2) - 1/(8a)$, $2 = 4\alpha^3\beta/(27b^2)$ and $3 = \alpha^3\beta/(18b^2)$

The LAOs are simply of relaxation type when, due to a "small" ε value, the system approximately traces the branches of $x_2 = a{x_1}^2 + b{x_1}^3 \equiv F(x_1)$ (the first equation in (2.2) for $\varepsilon = 0$) with negative slopes, including the neighborhoods of the maximum $(-2a/(3b), 4a^3/(27b^2))$ and minimum (0, 0) points of $F(x_1)$. The SAOs are a result of 2D-like oscillations in (x_1, x_2), see the graph in Fig. 2.4b projected into (x_1, x_2). In 3D the SAOs occur in the fold area of the (x_1, x_2, x_3) space around the origin (0, 0, 0). A series of SAOs around (0, 0, 0) is followed by one or more LAOs (relaxations). The system leaves the fold due to the canard explosion phenomenon to enter the relaxation mode. Then, the return mechanism after the relaxation phase, allows the system to return back to a neighborhood of the fold for a new series of SAOs in the next period. For example, in one period shown in Fig. 2.4b we have a series of 2 SAOs around the point (0, 0, 0) followed by 2 LAOs when the system follows two negative-slope branches of $F(x_1)$ with two (almost instantaneous) jumps between these branches. It was shown in [7] that the MMOs L^s happen if $c < \beta a^3/(18b^2)$. With the inequality sign reversed we only have LAOs and no SAOs. Special Stern–Brocot trees for (2.2) are constructed and discussed in [7], while analytical properties are considered in [8]. Finally, a numerical treatment of (2.2) is proposed in [9, 10], where a Fortran-90 code to solve the three-variable system, compute its bifurcation diagrams and Arnold's tongues are described.

Linearizing (2.2) yields the Jacobian matrix

$$J = \begin{pmatrix} \varepsilon^{-1}\left(2ax + 3bx^2\right) & -\varepsilon^{-1} & 0 \\ \alpha & -\alpha K & -\alpha \\ 0 & -\beta & 0 \end{pmatrix} \tag{2.4}$$

where $\alpha = L^{-1}$ and $\beta = C_2^{-1}$. Setting $\alpha = 1[\Omega \mathrm{s}]^{-1}$ and $K = 0$ we get the eigenvalues of (2.4) satisfying the characteristic equation

$$\lambda^3 + p\lambda^2 + q\lambda + r = 0 \tag{2.5}$$

where $p = (2ax + 3bx^2)\varepsilon^{-1}$, $q = \varepsilon^{-1} - b$, and $r = -(2ax + 3bx^2)\varepsilon^{-1}\mathrm{b}$. The Hopf locus is obtained for $r - pq = 0$, $q > 0$, which yields $(2ax + 3bx^2) = 0$. Thus from (2.5) we get $\lambda_{1,2} = \pm \mathrm{j}\sqrt{\varepsilon^{-1} - b}$, $\lambda_3 = 0$. Thus, the Hopf bifurcation occurs at $c = 0$. An example illustrating this problem is shown in Fig. 2.3a, with

$$\omega = Im\left(\lambda_{1,2}\right) = \sqrt{\varepsilon^{-1} - b} \cong 9.9997[\mathrm{rad/s}] \tag{2.6}$$

As for the roots of the Eq. (2.5) it is known [11] that it has a pair of complex conjugate solutions if the following condition is satisfied: $\Delta \equiv -4p^3r + p^2q^2 - 4q^3 + 18pqr - 27r^2 < 0$. For the MMOs the third solution of (2.5) must be negative. In the vicinity of the fold of the critical manifold we can approximate the system's dynamics by changing the coordinates

$$x_1 = \sqrt{\varepsilon}\tilde{x}_1,\ x_2 = \varepsilon\tilde{x}_2, x_3 = \sqrt{\varepsilon}\tilde{x}_3, t = \tilde{t}(\sqrt{\varepsilon})^{-1} \tag{2.7}$$

Substituting (2.7) into (2.2) yields

$$\begin{aligned}
\tilde{x}_1' &= -\tilde{x}_2 - a\tilde{x}_1^2 - b\sqrt{\varepsilon}\tilde{x}_1^2 \\
\tilde{x}_2' &= \tilde{x}_1 - \tilde{x}_3 \\
\tilde{x}_3' &= \varepsilon(c - \varepsilon\beta\tilde{x}_2)
\end{aligned} \tag{2.8}$$

where the prime now denotes differentiation with regard to the new rescaled time $\tilde{t}$, $a > 0$, $b < 0$, and $c > 0$ are O(1) coefficients, $0 < \varepsilon \ll 1$ is small, and β is the "free" (bifurcation) parameter; note the presence of three time-scales in (2.8).

It is the interplay between the two main ingredients of the dynamics, the local flow close to the strong canard and the global return, that underlies the basic canard mechanism for the emergence of MMOs in (2.2). Moreover, (2.2) is a normal form for the class of three time-scale circuits, in the sense that the addition of higher-order terms in it will not fundamentally influence the resulting dynamics. Another aspect of the mixed-mode dynamics in (2.2), in addition to the return mechanism, is the family of so-called secondary canards viewed as a trajectory that undergoes k small (nonrelaxation) rotations, or "loops," during its passage "near" the lower fold 1^- and that then remains $O(\varepsilon)$-close to the critical manifold S_0 until it reaches the $O(\varepsilon^{1/3})$-vicinity of the upper fold 1^+ [7, 12, 13].

Therefore, the generation MMOs in Chua's circuit requires the coordinated action of various mechanisms: (*i*) the generation of STOs; (*ii*) introduction of spiking action, including the onset of spikes and invoking of the spiking dynamics; and (*iii*) a return from the spiking regime back to the subthreshold regime [27, 28, 31, 32].

It is to be emphasized, that oscillations that exhibit mixed-mode-type behavior of a given circuit can be constructed from systems that undergo a canard explosion by replacing the parameter moving the slow nullcline with a dynamical variable. In other words, the emergence of MMOs in the Chua's type circuits is triggered by a "slow passage through a canard explosion." The term *canard explosion* is customarily used to express a transition in (2.2) from a stable equilibrium and subsequently to a large amplitude relaxation oscillation through a family of small-amplitude cycles [14]. The basic dynamics of a canard explosion can be interpreted as follows: (*i*) the "fast nullcline" S_0 for (2.2), which is given by $y = -f(x)\!: = ax^2 + bx^3$, is an *S*-shaped curve; (*ii*) S_0 is a curve of equilibria for the layer problem obtained for $\varepsilon = 0$ in (2.2) and is (normally) hyperbolic away from the two fold points where $f(x) = 0$; in particular, the origin is one such point; (*iii*) introducing in terms of ε the slow time $\tau = \varepsilon t$ in (2.2) one finds that the corresponding "slow nullcline" is given by $y = \lambda$ $(= c/\beta)$. As λ passes through 0, this slow nullcline moves through the lower fold point of S_0 at the origin, which triggers the onset of the canard explosion; finally, (*iv*) for $\lambda > 0$ sufficiently "large," the dynamics of (2.2) enters the relaxation regime.

2.5 Memristive Circuits with Steady-State Mixed-Mode Oscillations

For some years the memristor has been regarded as one of the most distinguished elements exhibiting both complex phenomena such as oscillation and chaos [12, 13, 16–19] and information processing with application in neural network [20–22], resistive switching memory [14], and artificial intelligence [8]. The growing exponentially interest on memristors is legitimated by their excellent features for both memory and neuromorphic applications [13, 16, 17]. For the former applications, they are nonvolatile and extremely small size devices of a few nano meters (Fig. 2.8a). For these second applications, they have features of pulse-based operations and controlable resistance (Fig. 2.8b), which are ideal for tuning the synaptic weights of neuromorphic cells. The circuit symbol of memristors is presented in Fig. 2.8c. In circuit theory memristors, as passive elements, complement the widely used other passive elements: resistors, capacitors, and inductors. It is characterized by the memristance function describing the charge-dependent rate of change of flux ϕ with charge q (Eq.(2.9)). Presently, each of the set of four passive two-port elements uses a pair of the current i, voltage v, charge q, or flux ϕ variables as their inputs and outputs (Fig. 2.9). In contrast to a linear (or non-linear) resistor the memristor has a dynamic relationship between current and voltage including a memory of past voltages or currents.

$$M(q) = \frac{\mathrm{d}\phi}{\mathrm{d}q} \tag{2.9}$$

Important features of memristors are both the non-linear voltage-current characteristic and memory because their present state at any instant depends on the past.

They all exhibit a peculiar "fingerprint" characterized by a pinched hysteresis loop confined to the first and the third quadrants of the v–i plane whose contour shape in general changes with both the amplitude and frequency of any periodic

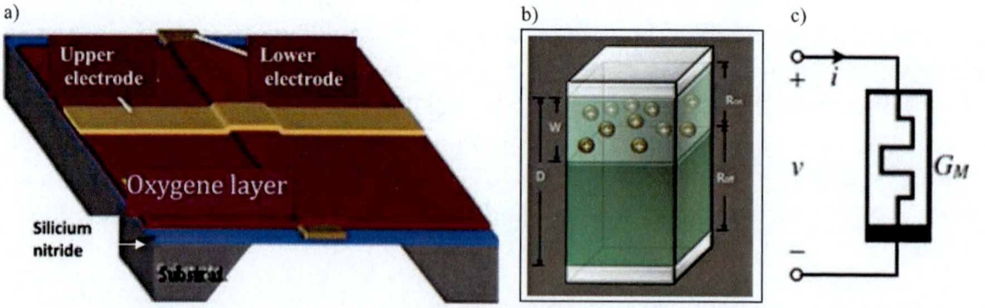

Fig. 2.8 Sketch of the memristor structure: **a** view of the real element, **b** conduction zones, **c** circuit symbol

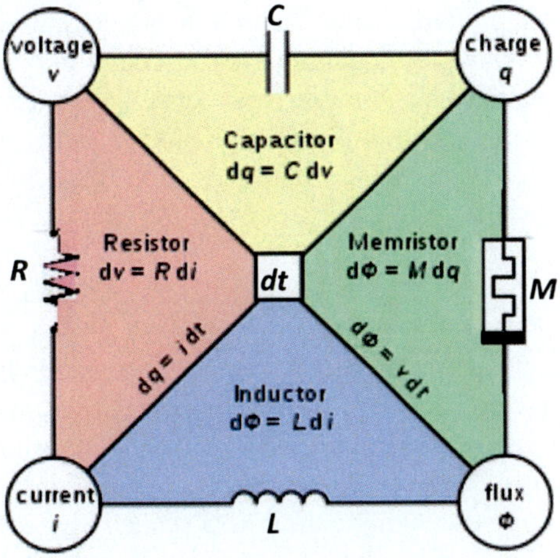

Fig. 2.9 Conceptual symmetries of resistor, capacitor, inductor, and memristor

"sinewave-like" input voltage source, or current source. At the origin of the v–i plane both signals are identically zero crossing. As frequency ω tends to ∞ the memristor degenerates into a linear R resistor. For example, the current-controlled voltage memristor is described by a relationship between the flux $\phi(t)$ and charge $q(t)$, as follows: $\phi(t) = F(q(t))$ with some function $F \in C^1$. This gives Ohm's law for such a memristor in the form: $u(t) = M(\int i(t)dt)i(t)$, with $M(q) = dF(q)/dq$ as the memristance, while $u(t)$ and $i(t)$ denote the voltage and current, respectively. The memory effect is due to the dependence of memristance M on $\int i(t)dt$.

The magnetic ***flux*** ϕ is generalized from the circuit characteristic of an inductor. It *does not* represent a magnetic field here. The symbol ϕ may be regarded as the integral of voltage $v(t)$ over time. Physically considering $M(q)$ is the positive-constrained parameter for all q-values (assuming the element is passive). An extra thermodynamic peculiarity arises from the definition that memristors should energetically act like resistors. The instantaneous electrical power entering such an element is completely dissipated as Joule heat to the surrounding, so no extra energy remains in the system after it has been brought from one resistance state x_n to another one x_m.

Memristor patents include applications in programmable logic, signal processing, neural networks, control systems, reconfiguration, brain-computer and RFID interfaces. Several recent work has been aimed at building an artificial brain, in which memristors can act as synapses that constitute connections between neurons.

In the general view of an n-th order memristive system the defining equations are

$$\begin{aligned} \dot{\boldsymbol{x}} &= \boldsymbol{f}(\boldsymbol{x}, \boldsymbol{u}, t) \\ y(t) &= \boldsymbol{g}(\boldsymbol{x}, \boldsymbol{u}, t)\boldsymbol{u}(t) \end{aligned} \tag{2.10}$$

where $\boldsymbol{u}(t)$ is an input signal, $\boldsymbol{y}(t)$ is an output signal, the vector $\boldsymbol{x}$ represents a set of n state variables describing the system, and $\boldsymbol{g}$ and $\boldsymbol{f}$ are ***continuous functions***. We henceforth adopt the standard notation $\boldsymbol{x}$ to denote a state variable in mathematical system theory, where $\boldsymbol{x}$ may be a vector $\boldsymbol{x} = [x_1, x_2, \ldots, x_n]'$. This will be the case for many non-ideal memristors found in practice. One of the resulting properties of memristors and memristive systems is the existence of a pinched hysteresis effect. The change in slope of the pinched hysteresis curves demonstrates switching between different resistance states. At high frequencies, memristive theory predicts that the pinched hysteresis effect will degenerate, resulting in a straight line representative of a linear resistor.

Typical MMOs exhibited in a 3D circuit containing a memristor are shown in Fig. 2.10: (a) time-series response 6^1 and (b) pinched hysteresis loop v-i. Observe also that the pinched hysteresis loop is odd symmetric with respect to the plane origin. Research and applications of memristive circuits exhibiting mixed mode oscillations are under systematic development due to their great importance in many fields of human activity. Those interesting periodic oscillatory variables have been reported in resistive nonlinear circuits (usually having cubic current–voltage characteristics) and in various chemical, biological and astrophysical systems.

Consider the two dual circuits shown in Fig. 2.11. There is one component of small constant value $\varepsilon > 0$ (either capacitor or inductor) in each of the circuits making them singularly perturbed ones. Both circuits are described by the following system of equations:

$$\begin{aligned} \varepsilon x\prime &= -y - \left(a + 3\mathrm{bw}^2\right)x \\ y\prime &= \alpha(x - \mathrm{Ky} - z) \\ z\prime &= -\beta y \\ w\prime &= x \end{aligned} \tag{2.11}$$

with $w = \phi$ (flux) in Fig. 2.11a and $\mathrm{w} = q$ (charge) in Fig. 2.11b where the prime (′) denotes the time derivative, $0 < C_1 \equiv \varepsilon \ll 1$, $\alpha = 1/L$, $K = R$, $\beta = \gamma/C_2$ for the

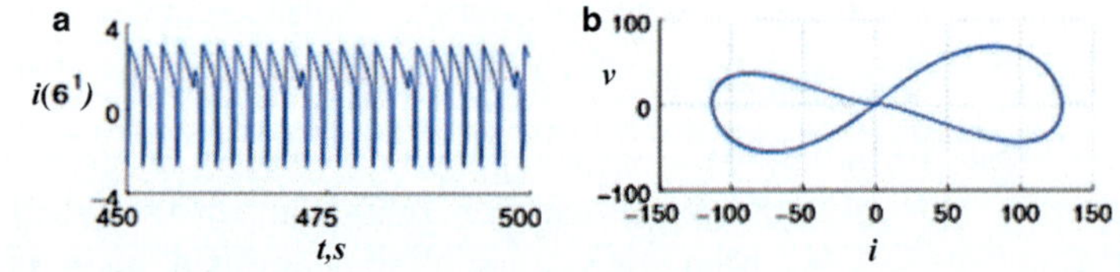

Fig. 2.10 MMOs in a memristive circuit: **a** slow-fast time-series 6^1, **b** pinched hysteresis $v - i$

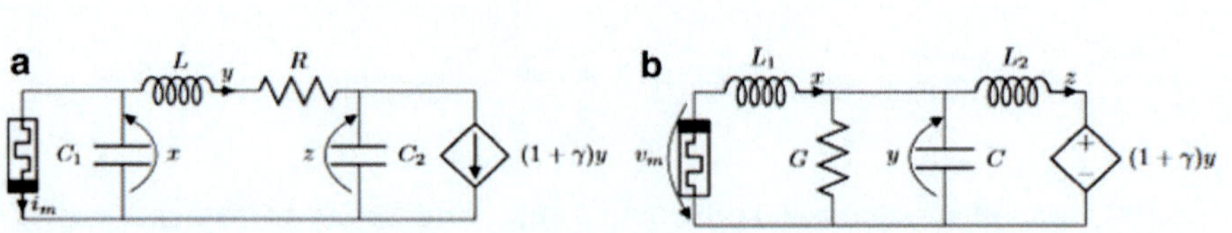

Fig. 2.11 Dual memristive circuits: **a** with current-controlled current source, **b** with voltage-controlled voltage source

circuit in Fig. 2.11a and $0 < L_1 \equiv \varepsilon \ll 1$, $\alpha = 1/C$, $K = G$, $\beta = \gamma/L_2$ for the circuit in Fig. 2.11b. The current-controlled current source (CCCS) and voltage-controlled voltage source (VCVS) in the circuits are described through expression $(1+\gamma)y$ with $\gamma > 0$. The small values of C_1 and L_1 in Fig. 2.11a, b make both circuits singularly perturbed ones. Suppose that we use the following parameters and initial conditions (ICs):

$$\begin{aligned} \varepsilon &= 0.01, \mathrm{K} = 0, \alpha = 1, \beta = 0.1, \\ x(0) &= 2.22,\ y(0) = 1.105, \\ z(0) &= -0.00628,\ w(0) = 0.37. \end{aligned} \tag{2.12}$$

For the purpose of illustration only the particular choice of the memductance $G(\phi(t)) = a + 3b\phi^2(\mathrm{t})$, a <0, $b > 0$, ϕ denoting the flux and parameter $\gamma > 0$ in (2.11) was chosen in the case of circuit with CCCS and VCVS.

Several typical steady-state MMO responses of the above circuits for various values of a and b (parameters of the memristive element) are shown below. First, Eq. (2.11) were solved with the variable step ode45 procedure from Matlab with relerr = abserr = 10^{-6} and $0 \le t \le 500$ s. The steady-state response and the large- and small-amplitude oscillations (LAOs and SAOs) were identified in the interval $450 \le t \le 500$ s. There is no significance attached to the chosen ICs.

The obtained MMOs are sensitive not only to the chosen ICs but also to the parameters in (2.11). For example, changing the b parameter from 0.356 (see MMOs 2^3) to 0.357 (with other parameters and ICs unchanged) results in MMOs 1^1. The s in 1^s indicates very large numbers of SAOs, all with very small amplitudes. Moreover, we have also observed the mode-locking phenomenon in the memristive circuits in Fig. 2.11, similar to those reported in the resistive MMO circuits in [17]. Various Arnold's tongues [1] can be computed in a two-dimensional parameter space, for example, in the $a - b$ rectangular area.

Figure 2.12a shows an example of the 2^3 MMOs response (variable $w(t) = q(t)$). Depending on the values L and s in L^s each period of MMOs can be divided into intervals of various natures: intervals with a sequence of SAOs and intervals of LAOs. One period of LAOs can be estimated as $T_L = T_1 + T_2$ (as shown in Fig. 2.12a). Note that the jumps between LAOs during T_1 and T_2 and also between SAOs within T_s and T_2 occur almost instantaneously and therefore can be ignored in the estimation of the period of MMOs. One SAO and one LAO oscillations have periods T_s and T_L, respectively, given by

$$T_s = 4\pi\sqrt{\frac{\varepsilon}{4a - K^2\alpha^2}} \tag{2.13}$$

$$\begin{aligned} T_L &= T_1 + T_2 = \int_{-w_0}^{-2w_0} \frac{\beta\left(a + 3bw^2\right)}{\alpha\left\{\beta w + K\left[\beta w\left(\alpha + bw^2\right) + B\right]\right\}} dw \\ &+ \int_{w_0}^{2w_0} \frac{\beta\left(a + 3bw^2\right)}{\alpha\left\{\beta w + K\left[\beta w\left(\alpha + bw^2\right) + B\right]\right\}} dw \end{aligned}$$

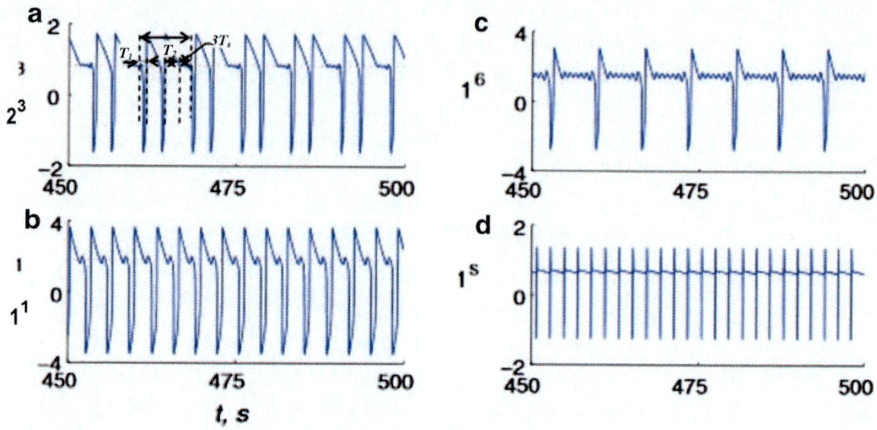

Fig. 2.12 MMOs: 2^3, 1^1, 1^6 and 1^s generated in circuit 2.11a

where $w_0 = \pm\sqrt{-a/(3b)}$ and $B = \mp(3 + 2aK)\sqrt{-a/(3b)}/(3\ \text{K})$. If a plus sign is used in w_0, then a minus sign is used in B and vice versa.

For 2^3 MMOs produced by circuit (2.11) and (2.12) we have two LAOs and three SAOs. Hence the period of the whole oscillation is

$$T = 2T_L + 3T_s \tag{2.14}$$

Taking into account $K = 1$, $\alpha = 1$, $\varepsilon = 0.01$, $a = -0.7$, $b = 0.35$, $\beta = 0.15$ and evaluating expressions (2.13) we get $T_s = 0.63$ s, $T_1 = 0.45$ s and $T_2 = 1.24$ s. Thus, in accord with (2.14) we have $T = 5.27$ s. Actual value of T estimated from the solution presented in Fig. 2.12a is 5.35 s.

There have been quite a number results reported on oscillatory memristor circuits in the literature recently [9, 10, 12, 13, 16–21, 23]. The pinched hysteresis loop for a single memristor and the area enclosed by a one-period loop have been studied in [13, 16, 20, 21], with the controlled memristor. When a memristor is a component of a complex nonlinear circuit, one may expect a much richer spectrum of relations between variables than in a simple situation when a single such element is considered only.

When all four variables v, ϕ, i and q for a memristor are taken into consideration and when it is a component of an oscillating circuit, a one-period 2D loops can be analyzed, as is shown through the numerical example with figures in what follows. The analysis of various types of closed loops for complex circuits with memristors must take into account not only the nonsinusoidal shapes of the involved signals, but also influences of other component parameters of the circuits. All memristors, regardless of the device material and physical operating mechanisms, are 2-terminal non-volatile memory devices based on resistance switching. In principle, a memristor can be used as a switch when an applied voltage will cause a significant change in its memristance. In such a switch, both the time and the energy necessary to achieve the desired change in memristance are the key qualitative factors [11, 24]. The switching of the memristor from M_{do} to M_{ndo} at the time T_1 to T_2 is accompanied by a change in the charge $\Delta q = q_{do} - q_{ndo}$. The energy consumption for such switching is quantified as $W_T = \int_0^T u(t)i(t)dt$ where T is the switching period. Thus, energy dissipation on switching takes place as with CMOS switches.

Taking into account the relationship between the changes in time of the flux $\phi(t)$ associated with the memristor or between the current and the changes in time of the charge q(t) carried by the memristor, we get:

- in the case of control by the flux

$$W = \int_{\psi(0)}^{\psi(T)} i(t)d\psi(t) \tag{2.15}$$

- in the case of control by the charge

$$W = \int_{q(0)}^{q(T)} u(t)dq(t) \tag{2.16}$$

The energy used to switch the memristor controlled by flux during the period of the voltage change is determined by the area bounded by a one-period energy loop (2.15) on the plane (i (t), ψ(t)). In the MMOs 2^3 example with $\varepsilon = 0.01$; $a = -0.7$; $b = 0.35$; $K = 0$; $\alpha = 1$; $\beta = 0.15$; $x_0 = [2.22; 1.105; -0.00628; 0.375]'$ we obtain the time series of the flux plot and the one-period energy loop presented in Fig. 2.13. Using the MATLAB *polyarea* function gives $W_T = 90{,}6612$ J for the numerically identified period $T = 5.3518$ s. Note that only a small area of the one-period energy loop corresponds to SAOs, which is related to the low amplitude of these oscillations.

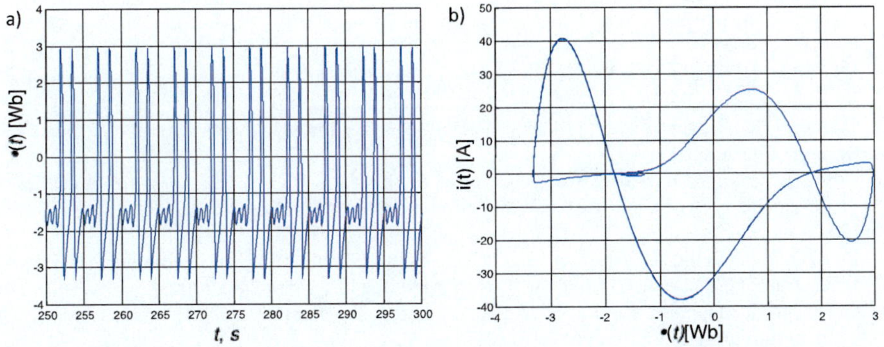

Fig. 2.13 Steady-state MMOs: **a** time series of the flux, **b** one period loop of energy

A similar relationship occurs when the duration of SAOs takes up a small part of the entire MMOs period. For example, after a slight change of the memristor control parameters and the remaining invariants, 6^1 MMOs are formed and the one-period energy reaches the value of 222.7570 J with the numerically identified period $T =$ 7.3518 s. In this case, the share of SAOs in the total single-period energy is even lower.

References

1. Trzaska, Z.: Dynamical processes in sequential-bipolar pulse sources supplying nonlinear loads. Electr. Rev. **90**(3), 147–152
2. Tharp, B.Z., Erdmann, D.B., Matyas, M.L., McNeel, R.L., Moreno, N.P.: The Science of the Heart and Circulation. Baylor College of Medicine, Houston (2009)
3. Rowell, L.B.: The cardiovascular System. In: Tipton, C.M. (ed) Exercise Physiology. American Physiological Society, Rockville (2003)
4. Solinski, M., Gieraltowski, J.: Different Calculations of Mathematical Changes of the Heart Rhythm. Center of Mathematics Applications, Warsaw University of Technology, Warsaw (2015)
5. Heldt, T., Verghese, G.C., Mark, R.G.: Mathematical modeling of physiological systems. In: Batzel, J.J., Bachar, M., Kappel, F. (eds) Included in Mathematical Modeling and Validation in Physiology: Applications to the Cardiovascular and Respiratory Systems. Springer, Berlin (2013)
6. Marszalek, W.: Bifurcations and Newtonian properties of Chua's circuits with memristors. DeVry Univ. J. Sch. Res. **2**(2), 13–21 (2015)
7. Subramaniam, B., Khabbaz, K.R., Heldt, T., Lerner, A.B., Mittleman, M.A., Davis, R.B., Goldberger, A.L., Costa, M.D.: Blood pressure variability: can nonlinear dynamics enhance risk assessment during cardiovascular surgery. J. Cardiothorac. Vasc. Anesth. **28**(2), 392–397 (2014)
8. MacEwen, C., Sutherland, Sh, Pugh, D.J., Tarassenko, L.: Validation of model flow estimates of cardiac output in hemodialysis patients. Ther. Apher. Dial. Off. Peer-Rev. J. Int. Soc. Apher. Jpn. Soc. Apher. Jpn. Soc. Dial. Ther. **22**(4), 1 (2018)
9. Lu, Z., Mukkamala, R.: Continuous cardiac output monitoring in humans by invasive and non invasive peripheral blood pressure waveform analysis. J. Appl. Physiol. **101**(2), 598–608 (2006)
10. Mukkamala, R., Reisner, A., Hojman, H., Mark, R., Cohen, R.: Continuous cardiac output monitoring by peripheral blood pressure waveform analysis. IEEE Trans. Biomed. Eng. **53**(3), 459–467 (2006)
11. Strasz, A., Niewiadomski, W., Skupi ́nska, M., Gąsiorowska, A., Laskowska, D., Leonarcik, R., Cybulski, G.: Systolic time intervals detection and analysis of polyphysiographic signals. In: Jablonski, R., Brezina, T. (eds) Mechatronics Recent Technological and Scientific Advances. Springer, Berlin (2011)
12. Heldt, T., Chang, J.L., Chen, J.J.S., Verghese, G.C., Mark, R.G.: Cycle-averaged dynamics of a periodically-driven, closed loop circulation model. Control Eng. Pract. **13**(9), 1163–1171 (2005)
13. Parlikar, T.A., Heldt, T., Verghese, G.C.: Cycle-averaged models of cardiovascular dynamics. IEEE Trans. Circuits Syst. I Fundam. Theory Appl. **53**(11), 2459–2468 (2006)
14. Genecand, L., Dupuis-Lozeron, E., Adler, D., Lador, F.: Determination of cardiac output: a game of thrones. Respiration **96**(6), 1 (2018)
15. Trzaska, Z.: Properties and applications of memristors—memristor circuits with innovation in electronics. In: Czy ̇z, Z., Maciąg, K. (eds) Contemporary Problems of Electrical Engineering

and Development and Evaluation of Technological Processes, pp. 76–93. Publisher TYGIEL, Lublin (2017)
16. Shibao, C., Grijalva, C.G., Raj, S.R., Biaggioni, I., Griffin, M.R.: Orthostatic hypotension-related hospitalizations in the United States. Am. J. Med. **120**, 975–980 (2007)
17. Fedorowski, A., Stavenow, L., Hedblad, B., Berglund, G., Nilsson, P.M., Melander, O.: Orthostatic hypotension predicts all-cause mortality and coronary events in middle-aged individuals. Eur. Heart J. **31**, 85–91 (2010)
18. Kozlovskaya, I.B., Grigoriev, A.I.: Russian system of countermeasures on board of the International Space Station (ISS): the first results. Acta Astronaut. **55**, 233–237 (2004). https://doi.org/10.1016/j.actaastro.2004.05.049
19. Walchenbach, R., Geiger, E., Thomeer, R.T., Vanneste, J.A.: The value of temporary external lumbar CSF drainage in predicting the outcome of shunting on normal pressure hydrocephalus. J. Neurol. Neurosurg. Psychiatry **72**, 503–536 (2002)
20. Arnold, A.C., Raj, S.R.: Orthostatic hypotension: a practical approach to investigation and management. Can. J. Cardiol. **33**, 1725–1728 (2017)
21. Low, Ph.A., Tomalia, V.A.: Orthostatic hypotension: mechanisms, causes, management. J. Clin. Neurol. **11**(3), 220–226 (2015). https://doi.org/10.3988/jcn.2015.11.3.220
22. Adamec, I., Junakovic, A., Krbot, M., Habek, M.: Association of autonomic nervous system abnormalities on head-up tilt table test with joint hypermobility. Eur. Neurol. **79**(5–6), 319–324 (2018)
23. Izhikevich, E.M.: Dynamical Systems in Neuroscience: The Geometry of Excitability and Bursting. MIT Press, Cambridge (2007)
24. Rajzer,M., Kawecka-Jaszcz, K.: Arterial compliance in arterial hypertension. From pathophysiology to clinical relevance. Arter. Hypertens. **6**(1), 61–73 (2002)
25. Niewiadomski, W., Gasiorowska, A., Krauss, B., Mróz, A., Cybulski, G.: Suppression of heart rate variability after supramaximal exertion. Clin. Physiol. Funct. Imaging **27**(5), 309–319 (2007)
26. Dimitrakopoulos, E.G.: Nonsmooth analysis of the impact between successive skew bridge-segments. Nonlinear Dyn. **74**, 911–928 (2013)
27. Trzaska, Z.: Nonsmooth analysis of the pulse pressured infusion fluid flow. Nonlinear Dyn. **78**, 525–540 (2014)
28. Lesniak, M.S., Clatterbuck, R.E., Rigamonti, D., Williams, M.A.: Low pressure hydrocephalus and ventriculomegaly: hysteresis, non-linear dynamics, and the benefits of CSF diversion. Br. J. Neurosurg. **16**, 555–561 (2002)
29. Giusti, A., Mainardi, F.: A dynamic viscoelastic analogy for fluid-filled elastic tubes. Meccanica **51**(10), 2321–2330 (2016)
30. Colombaro, I., Giusti, A., Mainardi, F.: On the propagation of transient waves in a viscoelastic Bessel medium. Z. Angew. Math. Phys. **68**(3), 62 (2017)
31. Marszalek, W.: Fold points and singularity induced bifurcation in inviscid transonic flow. Phys. Lett. A **376**, 2032–2037 (2012)
32. Podhaisky, H., Marszalek, W.: Bifurcations and synchronization of singularly perturbed oscillators: an application case study. Nonlinear Dyn. **69**, 949–959 (2012)
33. Wolf, A., Swift, J.B., Swinney, H.L., Vastano, J.A.: Determining Lyapunov exponent from a time series. Physica **16D**, 285–317 (1985)

Chapter 3
MMOs in Biological Processes

3.1 Introduction

Biosystems are extremely myriad natural systems of great diversity. All biosystems are extremely complex, i.e. highly organized, polyphasic and store enormous amounts of information. The idea of understanding how the biosystem is working has fascinated almost all societies for many centuries. Nowadays, thanks to achievements of informatics and modern scanning techniques, it has become possible to generate superb images of such structural components and processes in biosystems as neurons and fibers, blood flow and energy metabolism, and changes in neuronal activity. Advanced information systems, such as living cells, provide the information needed to transform incoming energy into useful work—transformation under certain circumstances impossible in the case of physical systems that lack comparable levels of organization. Automation and miniaturisation are particularly important at the interfaces between chemistry, biotechnology, biology and medicine and it is likely that the rise of genomic and post genomic technologies will drive the need for chemical and biochemical reactions on surfaces. There have been many successful applications in diagnostics, omics studies, sensors, plant sciences, enzyme kinetics, enzyme specificity, cell biology, and organic (synthetic) chemistry [1–11]. Many biological systems consist of networks of coupled oscillators generating a large multiplicity of widespread behaviors that include periodic, quasiperiodic, and aperiodic rhythms. They putatively underlie empirical phenomena such as physiological and pathophysiological oscillations, cardiac arrhythmia, fluid turbulence, spiral waves and cluster patterns in excitable media, and deterministic chaos in many forms. Using principles from chemistry, physics and mathematics with applications of informatics, it can be seen that the highly complex behavior of biological systems is caused by a multitude of coupled feedback and feed-forward loops in the underlying auto control neuronal networks.

Realized up to date biological research has accumulated a powerful amount of detailed and inestimable knowledge about the structure and function of the brain. The human brain is the most complex organ of the body and probably the most

Z. Trzaska, *Mixed Mode Oscillations (MMOs)*, Studies in Systems, Decision and Control 374, https://doi.org/10.1007/978-3-030-76867-6_3

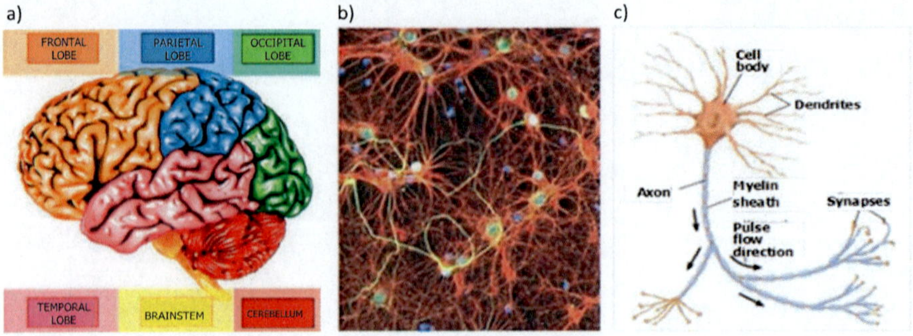

Fig. 3.1 Components of the human central nervous system: **a** brain, **b** bit of neural network, **c** single neuron

complicated system in the nature. It is a unrepeatable organ that thinks and feels, produces behavioral interactions with the environment, maintains bodily physiology relatively stable, and enables regeneration of the species being its most important role from evolution's impressive perspective, capable of exchanging genes or interbreeding. The elementary active units in the central nervous system are neurons, which are connected to each other in an entangled complex (Fig. 3.1a). A neuron network diagram (Fig. 3.1b) of cells (blue) and axons (red), with axon extent indicating its strength, portrays the information streamline density linking each cell pair. For clarity, only the strongest axons are shown in Fig. 3.1b.

Neurons are formed up of a cell body connected with two types of protrusions: axons transferring stimuli and dendrites receiving information (Fig. 3.1c). An important role in passing and getting impulses between neurons play special connections called synapses. There are three main types of neurons: sensory neurons, motor neurons and interneurons. Sensory neurons are linked to receptors responsible for getting and reply to various internal and external environment stimuli. They are sensitive to changes in lighting, sound, and they support mechanical and chemical stimulation the senses of sight, hearing, touch, smell and taste. Motor neurons control muscle activity, and are responsible for all forms of behavior, including speech. The activities of interneurons are located between those that characterize sensory and motor neurons. Their role lies in connection two brain regions, i.e. not direct motor neurons or sensory neurons. Interneurons are the central nodes of neural networks, enabling communication between sensory or motor neurons and the central nervous system. They mediate in the transmission of the simplest stimuli, but they are also responsible for the most complex brain functions. Support glial cells have a great influence on the development of the nervous system as well adult brain functioning. They are a lot more numerous than neurons, but they do not conduct stimuli in a characteristic way for the former ones.

The central nervous system, consisting of the brain and spinal cord, functions to receive, interpret, and respond to the nerve impulses of the peripheral nervous system. The peripheral nervous system serves as the connection between the central nervous system and the body, transporting sensory and motor information to and from the central nervous system, respectively. Neurons are organized into complex chains and networks, which are the routes of transmitting information in the nervous system. Each neuron can connect up to 10,000 other neurons. The entire nervous network is in a state of continuous electrical and chemical activities.

Approximately mentioning, the nervous system is built mainly from nerve cells called neurons and glial cells as supporting cells, including Schwann cells, satellite cells, oligodendrocytes, microglia, astrocytes, and ependymal cell. The human brain is made of neural networks, contains approximately 100 billion nerve cells, 3.2 million kilometers of "wiring" formed by axons and dendrites, one million billion connections (synapses), and a volume of 1.5 L, weighs about 1.5 kg, and works with power only 10 W. Understanding the development and homeostasis of the nervous system, the many factors responsible for maintaining and regulating its proper functionality, and the implications of nervous system abnormality, remain most important within research concerning the nervous system and neurodegenerative diseases. Research aimed at the biological processes of the nervous system and their influences on both human behavior and activity offers extensive possibilities for the comprehension as well as treatment, of various neurological diseases and ailments, such as Alzheimer's disease, multiple sclerosis, Parkinson's disease and epilepsy.

The idea of understanding how the brain is working has fascinated almost all societies for many centuries [12–15].

Mathematical modeling has a great potential for increasing our understanding of the physiological processes underlying the function of the human body. For example, modeling of the electrical properties of excitable neuron cells may provide insight into the complex electrical signaling revealed in a number of important activities, like transfer of information through neurons and coordination of the cardiovascular system.

Oscillations in neuronal networks have been of great interest to neuroscientists for many decades. They arguably represent the simplest form of coherent network behavior in the brain. One of the most interesting facts about brain oscillations is that they provide a possible mechanism by which Hebbian cell assemblies, thought to be the basic unit of information storage in the brain, can be created and held together. The so-called Hebbian cell assemblies are produced, in accordance with the well-known hypothesis of synaptic plasticity and memory [16–18], by self connecting neurons that strengthen synapses depending on their neuronal activity [16]. These sets of cells are believed to form the basis of long-term memory. It should be emphasized that models as simple as those discussed in this chapter obviously don't come close to reproducing the complexities of real human brain. The complexity of the brain is astonishing, and we currently cannot follow it faithfully in mathematical models and computational simulations. The presentation of the biological background in this chapter is therefore highly selective and focuses on those aspects that are needed to highlight the biological background of the theoretical work presented in this book. For an in-depth discussion of neurobiology, the reader is referred to the literature listed at the end of the book chapter. It is worth noting that many functions of the brain are simply not well understood experimentally, and computers are not (yet) powerful enough to reveal all significant problems. But, during the last 50 years computational thinking has importantly improved our understanding of biological systems. Moreover, in recent years major steps have been taken to understand biological algorithms by synthesizing biological regulatory networks de novo, which aim to compute specific functions [33, 35, 38].

3.2 The Hodgkin–Huxley Neuron with MMOs

In biological systems interesting phenomena take place close to the neuron cell membrane. Neurons are highly polarized cells, meaning that they form distinct subcellular zones that perform different functions. From morphological point of view, a typical neuron can be divided in three major parts: (1) the cell body containing the nucleus and the major cytoplasmic organelles; (2) a tree of dendrites, which originate from the perikaryon and ramify over a certain region of gray matter and which differ in size and shape, depending on the neuronal type; and (3) a single axon, which, in most cases, takes a form of a truncated cone extended more farther from the cell body than the dendritic tree (Fig. 3.2). Neurons and other cells are filled and surrounded by water in which ions such as sodium (Na^+), potassium (K^+), chloride (Cl^-), and calcium (Ca^{2+}) are dissolved. The superscripts indicate electrical charge, i.e., the charge of a positron: $q \approx 1.60 \times 10^{-19}$ C. The charge of an electron is $-q$. Most ion species can pass from the interior of the cell (ICF) to its exterior (ECF) or vice versa through ion channels in the cell membranes. Often these channels are specific: sodium channels only let sodium pass, potassium channels only let potassium pass, and so on. Since ions can pass from one side of the membrane to the other, one might expect their concentrations to become the same on both sides. However, this is not what happens because of two complications. First, some ions are not capable of crossing the membrane. There are, in particular, large negatively charged molecules present only in the cell interior, not in the extracellular fluid. Second, cell membranes contain pumps that actively transport ions from one side to the other. The most famous of these is the sodium–potassium pump, which removes

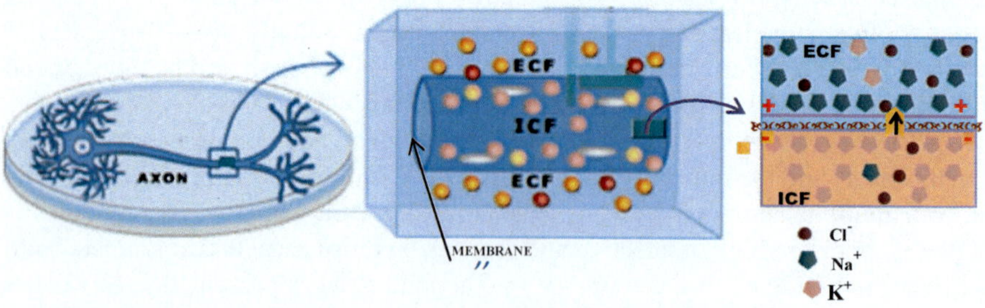

Fig. 3.2 Scheme of components of nervous systems

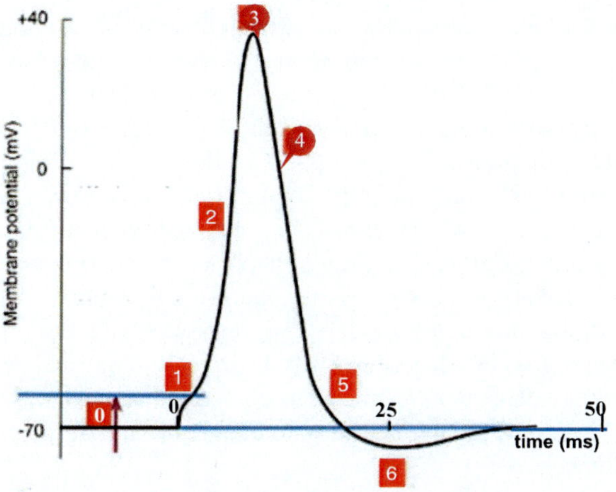

Fig. 3.3 Time-varying action potential: ⓪—threshold of excitation; ① Na^+ channels open, Na^+ begins to enter cell; ② K^+ channels open, K^+ begins to leave cell; ③ Na^+ channels become refractory, no more Na^+ enters cell; ④ K^+ continues to leave cell, causes membrane potential to return to resting level; ⑤ K^+ channels close, Na^+ channels reset to enter cell; ⑥ extra K^+ outside diffusive away

sodium from the cell and brings in potassium—2 potassium ions for 3 sodium ions. Suffice it to say that in equilibrium, ion concentrations in the extracellular fluid are different from those in the cell interior, and there is an electrical potential jump across the cell membrane. The difference in electrical potential between the interior and the exterior of the membrane is called the membrane potential, denoted by v. For a nerve cell in equilibrium, a typical value of v might be -70 mV, i.e., the potential on the interior side of the membrane is 70 mV below that in the extracellular fluid (Fig. 3.3).

From 1938 to the publication in 1952 of five landmark, among seven papers published between 1949 and 1955 A. L. Hodgkin and A. F. Huxley with their collaborators exposed the results of many of their works, both theoretical and experimental, in which they attempted to explain the basic physical mechanism by which, in humans and animals, electrical impulses are generated by nerve cells, and transmitted along axons. They clarified the mechanism underlying the action potentials in the space-clamped squid axon, and summarized this action in the form of a system of ordinary differential equations (ODEs). For their discoveries concerning the ionic mechanisms involved in excitation and inhibition in the peripheral and central portions of the nerve cell membrane they, together with J. C. Eccles, won the 1963 Nobel Prize in Physiology and Medicine. The Hodgkin-Huxley theory allowed the foundation for modern computational neuroscience providing the adequate model system. Viewed as a neuron model with concentrated parameters it describes the way a neuron deals with an input voltage to produce or not to produce an action potential. However, such models do not address the more complex features of neurons that may influence the formation of networks and dissemination of an action potential. Therefore, the Hodgkin-Huxley model simulates only the biological functioning of a neuron. Different parameters of the model represent specific biological components of a neuron.

The morphology of the axon and its course through the nervous system are correlated with the type of information processed by the particular neuron and by its connectivity patterns with other neurons. At the interface of axon terminals with target cells are the synapses, which represent specialized zones of contact consisting of a presynaptic (axonal) element, a narrow synaptic cleft, and a postsynaptic element on a dendrite or perikaryon containing the nucleus and a variety of cytoplasmic organelles. The Hodgkin-Huxley model [4] is based on the assumption that the membrane contains proteins that selectively conduct sodium and potassium ions in a time- and voltage-dependent manner. The elaborated model focuses particularly on the electrical attributes of the neural membranes during signal generation and propagation in the nerves.

When a current enters the cell it changes the membrane potential. An action potential, the rapid movement of ions, starts when the membrane potential reaches a prescribed threshold due to external sources of input current. When the membrane potential reaches said threshold, Na^+ channels open allowing Na^+ to enter the cell (flood the intracellular fluid) which causes the membrane potential to spike. K^+ also have channels to allow K^+ ions to leave the intracellular fluid but these channels require a higher membrane potential voltage than the Na^+ channels. Therefore, first Na^+ enters the cell and creates a sharp increase in the membrane potential. This increase then causes the K^+ channels to open, allowing K^+ to leave the intracellular fluid. At this point, Na^+ is entering faster than K^+ is leaving because of the electrostatic and diffusion forces exerted on it, so there is a still a net gain in membrane potential. When the peak occurs, Na^+ channels become refractory (blocked) so Na^+ can no longer enter the cell, but K^+ can still leave, thereby making the membrane potential decrease (Fig. 3.4). Once the resting potential of the membrane is reached K^+ channels close and Na^+ channels reset so that they can eventually be opened again. At the end of the action potential, the mix of Na^+ and K^+ in the intra- and extracellular fluid is not the stable electrolytic solution. Therefore the forces of diffusion and electrostatic pressure are not balanced, producing a refractory period where the cell membrane goes below the resting point before increasing again as the ions move into their stable states. Ion movements and the subsequent action potentials form a background of the neurons working.

The Hodgkin-Huxley model treats the nerve axon as an electrical circuit in which the proteins are resistors and the membrane is a capacitor. Ion currents flow through the membrane and along the nerve axon leading to a propagating pulse.

Assuming that a nerve cell membrane acts like a capacitor, separating two charge layers of opposite signs, it is possible to derive the equation describing how the action potential v evolves with time. It equals to a jump in the electrical potential across the cell membrane with the separated charges $\pm$ q.

The equation governing the instantaneous state of the capacitor has the form

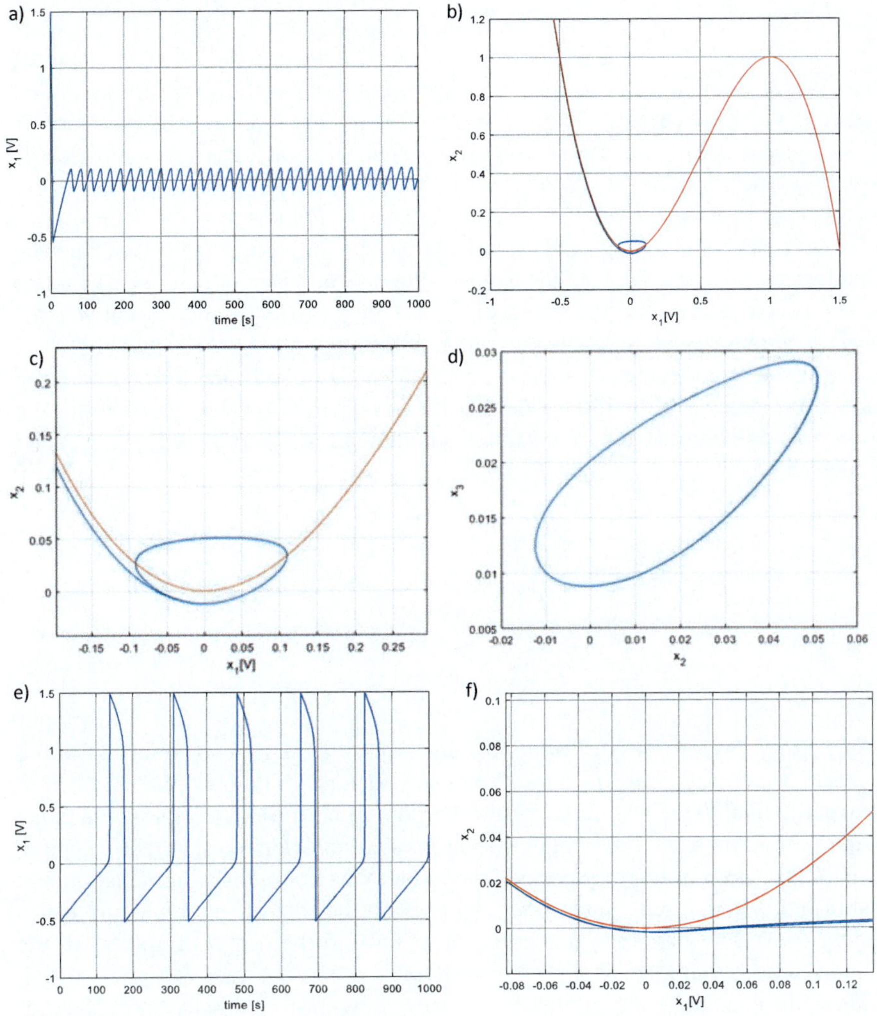

Fig. 3.4 Illustration of solutions of (3.3) for different neuron parameters: **a** spikes, **b** phase space $x_2(x_1)$, **c** transition to STO, **d** phase space $x_3(x_2)$, **e** relaxation oscillation, **f** phase space $x_2(x_1)$

$$C\frac{dv}{dt} = i_{tot} \tag{3.1}$$

where C is the membrane capacitance and i_{tot} denotes the total current.

Three different ion currents contribute to the voltage signal of the neuron, that is, a sodium current, a potassium current, and a leak current that consists mainly of Cl^- ions. The flow of these ions through the cell membrane is controlled by their respective voltage-dependent ion channels.

The total current is given by

$$i_{tot} = -(i_K + i_{Na} + i_{leak} + i_{stim}) \tag{3.2}$$

where i_K, i_{Na}, i_l and i_{stim} are, respectively, the potassium ion current, the sodium ion current, the leakage current and the stimulus current. The i_K component is given by $g_K h^4(v, t)(v - E_K)$ $dh(v, t)/dt = -(\alpha_h + \beta_h)h(v, t) + \alpha_h$, The i_{Na} component is given by $g_{Na} m^3(v, t)(v - E_{Na})$ with, $dm(v, t)/dt = -(\alpha_m + \beta_m)m(v, t) + \alpha_m$, where α_p and β_p, $p = h, m$ are voltage dependent. The i_L component given by $g_{leak}(v - E_{leak})$ with $g_{leak} =$ const. is small value and will be omitted in the following. The quantities E_{Na}, E_K and E_{leak} are the Nernst potentials of different ions [1]. The terms g_{Na} and g_K are the conductances of the ion channel proteins for the respective ions. They are assumed to be functions of voltage and time.

Taking into account the functions m and h called gating variables ranging between 0 and 1 and related to the likelihood that the ion channel is open which fulfill the above simple linear differential equations and combining them with Eqs. (3.1) and (3.2) yields

$$\begin{aligned} \frac{dv}{dt} &= f_1(v,m,h,t) \\ \frac{dm}{dt} &= f_2(v,m,h,t) \\ \frac{dh}{dt} &= f_3(v,m,h,t) \end{aligned} \tag{3.3}$$

where $f_1(\bullet)$, $f_2(\bullet)$ and $f_3(\bullet)$ are nonlinear functions of the action potential and gating variables. Their explicit forms depend on many factors, and most often are fitted to the experimental data. The model could reproduce and explain a remarkable range of data from squid axon, including the shape and propagation of the action potential, its sharp threshold, refractory period, anode-break excitation, accommodation and subthreshold oscillations. Such model focuses on the role of the ionic currents in shaping the MMO patterns (generation of STOs and the origin of spikes) or the dynamic actions that give evolution to MMOs.

Most of the neurons are excitable, i.e., they show quiescent behaviour, however when they are stimulated by input impulse they can also fire spikes. The complexity of spikes or their trains can be controlled by external stimuli, e.g. by injected electrical currents.

Very often the neurons continue to fire a train of spikes when there is an input stimulus by injecting a pulse of current and this is called tonic spiking. There exist different types of spiking patterns depending on the nature of the intrinsic dynamics of the neuronal network. One of the interesting complex firing patterns emerge from the activity of neurons is the mixed-mode oscillations (MMOs) [15–17], which are a combination of small amplitude (sub-threshold) oscillations interspersed with large amplitude relaxation type oscillations what is the main focus of this section.

The generation of MMOs requires the coordinated action of various mechanisms: (i) a mechanism for the generation of subthreshold oscillations (STOs); (ii) a spiking mechanism, including the onset of spikes and the description of the spiking dynamics;

and (iii) a return mechanism from the spiking regime back to the subthreshold regime. It should be emphasized that phenomena realized in neurons can be depicted as the interaction of rudimentary subprocesses. It can be also stated that very often different subprocesses evolve at distinct time scales. It is known that the separation of variables changing with different time scales may introduce interesting and complicated system dynamics, including exceptional responses known as canards. In the case where there is just one slow variable in a system, then canard responses appear as explosive growths in oscillation amplitude of a response to a very small change in system parameters. However, when in a system there are two slow variables the existence of canard responses can lead to complicated oscillatory patterns in the form of MMOs [30–32].

Applying the model (3.2) with $f_1(\bullet) = 3v^2 - 2v^3 - h$; $f_2(\bullet) = 0.008v - 0.002\,h - 0.003\,m$ and $f_3(\bullet) = 0.02v - 0.0135\,m + 0.00005$ gives MMO patterns presented in Fig. 3.5 for $x_1 = v(t)$, $x_2 = h(t)$ and $x_3 = m(t)$, where v is the membrane potential (mV), C is the membrane capacitance (μF/cm^2), Iapp is the applied bias (DC) current (μA/cm2), $I_{Na} = G_{Na}m^3h\,(V - E_{Na})$, $IK = GK\ \mathrm{n}^4\ (\mathrm{V} - \mathrm{E_k})$, $I_L = G_L\,(V - E_L)$, $I_{Nap} = Gp\ p_\infty(V)\,(V - E_{Na})$, $Ih = Gh\,(cf \bullet rf + \mathrm{cs} \bullet rs)\,(V - E_h)$. The parameters G_X and E_X ($X = Na, K, L, p, h$) are the maximal conductances (mS/cm^2) and reversal potentials (mV), respectively. The unit of time is *ms*. The variables *rf* and *rs* are the *h*-current fast and slow gating variables and the parameters *cf* and *cs* represent the fraction of the total *h*-current corresponding to its fast and slow components respectively.

The values of the parameters have been used [19, 20]: $E_{Na} = 55$, $E_K = -90$, $E_L = -65$, $E_h = -20$, $G_{Na} = 52$, $G_K = 11$, $G_L = 0.5$, $G_p = 0.5$, $G_h = 1.5$ and $C = 1$.

In the diagrams presented in Fig. 3.5 all the MMO mechanisms are described by the same model.

The generation of STOs (Fig. 3.4a) and the onset of spikes (Fig. 3.4e) are governed by the locally parabolic portion of the v-nullcline, near its minimum. The spiking dynamics and the return mechanisms are primarily governed by the right and left branches of the v-nullcline respectively (Fig. 3.4b). The abrupt transition from STOs to spikes is due to the 2D (Fig. 3.5a, b) and 3D canard phenomena (Fig. 3.5c) [21]. The locally parabolic nonlinearity at the minimum of the v-nullcline and the time scale separation between v and the remaining dependent variables are key for these mechanisms [36, 37].

There have been various attempts to simplify the complexity of the equation, for instance by FitzHugh and Nagumo [22, 23] (called the FitzHugh–Nagumo (FHN) model), Hindmarsh–Rose [24] and Rajagopal [21]. Such models are widely used in simulating the neural networks in order to rationalize experimental data.

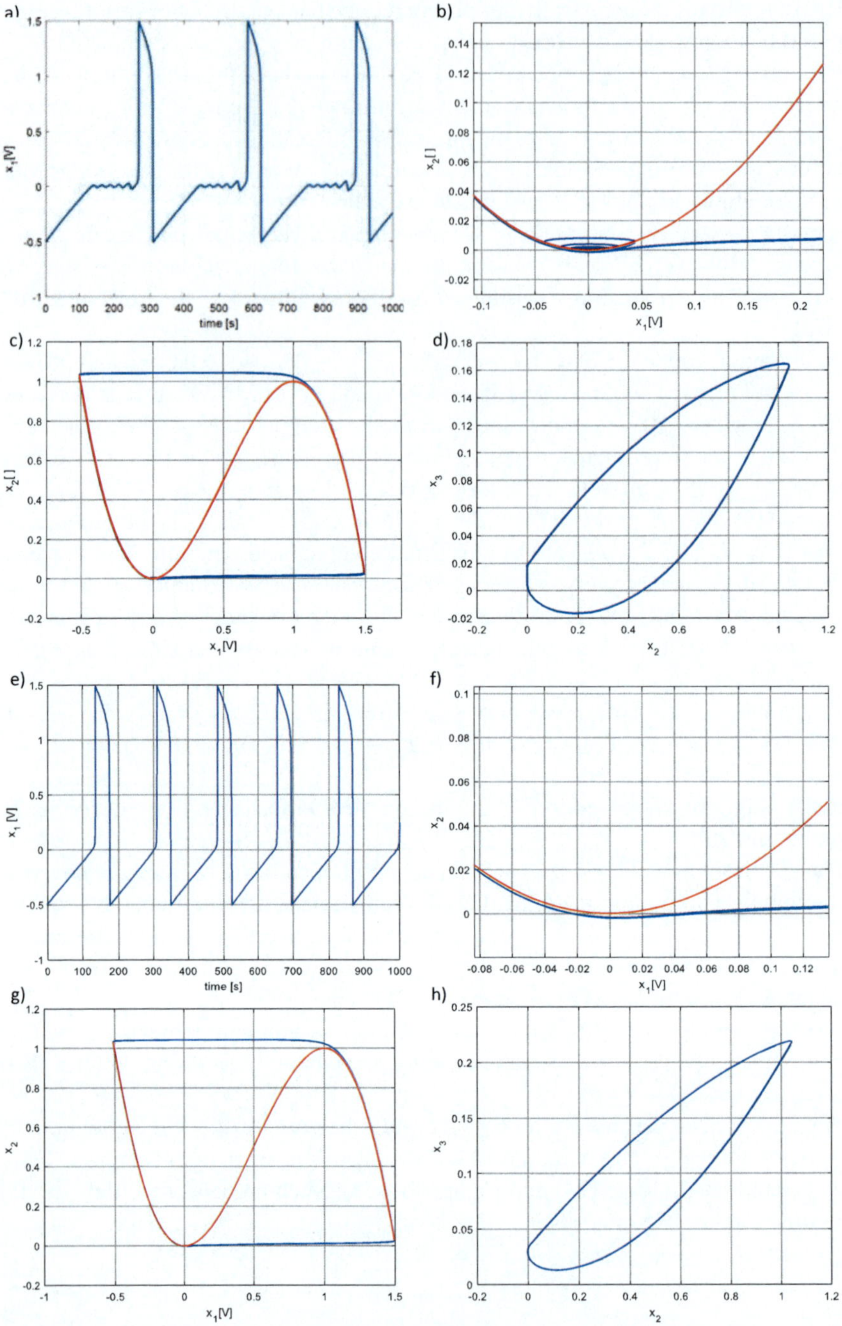

Fig. 3.5 MMOs with (3.3) for different neuron parameters: **a** 1^5 MMOs, **b** nullcline $x_2(x_1)$ and phase portrait for SAOs, **c** transition to STO, **d** trajectory in phase space $x_3(x_2)$, **e** relaxation oscillation, **f** trajectory in phase space $x_2(x_1)$, **g** transition to STO, **h** trajectory in phase space $x_3(x_2)$

3.3 Reduced Model of a Single Neuron Activity

Since the seminal work of Hodgkin and Huxley, there has been various attempts to simplify the complexity in the dynamical systems view point of a neuron, for instance by FitzHugh and Nagumo [22, 23].

The FHN model is frequently used as a reduced two-dimensional model for the description of neural oscillations and of other phenomena in fast–slow systems. Models of FHN type capture the basic dynamic properties observed in various conductance-based approaches to neuron activities. They represent the minimal models being able to generate intrinsic oscillatory neuronal behavior and involve the dynamics of the voltage v and a recovery variable w. In such a model, the self-coupling is provided by an extra synaptic variable whose evolution alternates between fast and slow modes. These models may generate either STOs (Fig. 3a) or spikes (Fig. 3.5b) but not MMOs.

Assuming similar dynamics of slow sodium and potassium ions and replacing $m(t)$, $h(t)$ in (3.3) by one effective current $w(t)$ leads to the FitzHugh-Nagumo model represented by the following two nonlinear ordinary differential equations

$$\frac{dv}{dt} = f(v, w)$$
$$\frac{dw}{dt} = g(v, w)$$

where v and w denote the action potential and recovery variable, respectively. The field flow functions $f(v,w)$ and $g(v,w)$ are determined by relatively low degree polynomials of their arguments.

In classical approach the basic FHN model takes the form

$$\begin{aligned} \frac{dv}{dt} &= v - \frac{1}{3}v^3 - w + i_{stim} \\ \frac{dw}{dt} &= \frac{av - bw}{\tau_w} \end{aligned} \tag{3.4}$$

where a, b, τ_w and i_{stim} denote the scaling parameters, time constant and the stimulus current, respectively. Usually $a > 0$, $b < 0$ and $\tau_w \gg 1$. For FitzHugh-Nagumo made the observation that when the Hodgkin/Huxley equations exhibit periodically firing action potentials, the following approximately holds:

$$n(t) + h(t) \approx 0.8. \tag{3.5}$$

Equation (3.5) should be viewed as an observation only—it has no rigorous mathematical or biological basis. However, it does allow us to further simplify the Hodgkin/Huxley model.

We choose to eliminate the gating variable h by taking $h(t) = 0.8 - n(t)$.

All variables and parameters should be considered dimensionless here. FitzHugh's choices, translated into our notation are: $a = 1.25$ and $\tau_w = 15.625$. The left panel

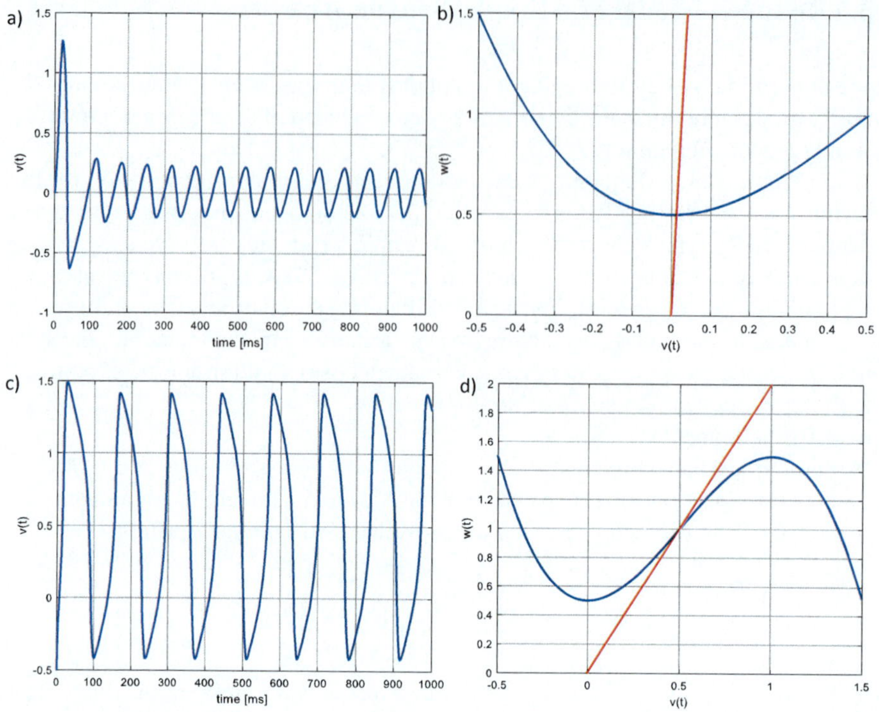

Fig. 3.6 Simulation results with (3.4) and (3.5): **a** spikes, **b** nullcline, **c** relaxation oscillation, **d**

of Fig. 3.6 is the analogue of Fig. 3.4a, but for the FitzHugh-Nagumo model, with $a = 1.25$ and $\tau_w = 25$. The right panel shows w as a function of v. In spite of its simplicity, the FitzHugh-Nagumo model can reproduce aspects of the dynamics of the classical Hodgkin-Huxley model surprisingly well.

The generation of STOs (Fig. 3.6a) and the onset of spikes (Fig. 3.6c) are governed by the locally parabolic portion of the v-nullcline, near its minimum. The spiking dynamics and the return mechanisms are primarily governed by the right and left branches of the v-nullcline respectively (Fig. 3.6b). The abrupt transition from STOs to spikes is due to the 2D (Fig. 3.6a, c) and 3D canard phenomena.

3.4 MMOs in a Nonlinear Human Cardiovascular System

3.4.1 Concise Characterization of a Cardiovascular System

One of the most important organs in human body is heart which pumps blood throughout the body using blood vessels. The heart works as two pumps acting simultaneously, one on the right and one on the left (Fig. 3.7). Together with blood and blood vessels the heart belongs to the human cardiovascular system. The functional and structural characteristics of the blood vessels change with successive branching. Thanks to the heart contraction, the blood flow is pulsatile and blood is pumped into the two circulations by means of discrete pulses with a pressure usually varying during one heartbeat in the ranges 70–130 mmHg and 20–30 mmHg for the systemic and pulmonary networks, respectively. The work of the heart is to pump blood to the lungs through pulmonary circulation and to the rest of the body through systemic circulation accomplished by systematic contraction and relaxation of the cardiac muscle in the myocardium. Normally, for the right ventricle, work is about 16% of the work done by the left ventricle [34].

Over the past several decades, important research efforts have been focused on investigations of cardiovascular system responses to changes caused by various diseases affecting malfunction in the heart to a point where the heart can no longer pump enough blood to the cardiovascular system [14–18]. Untreated diseases like coronary heart disease, valvular disease, or hypertension have important influences on heart malfunction through different factors that lead eventually to potential clinical impacts. To determine the factors causing the insufficient functioning of the cardiovascular system and its deficiencies, one can reach the significant help in the form of mathematical modeling allowing to isolate the contribution of several key parameters from the whole cardiovascular process, and trying to predict the dominant factors, as well as changes in the patient state. In this way, clinical examinations can be minimized, especially at the expensive stage of the pilot investigations, and then the costs of developing a practical procedure of treatment combination of cardiotherapy are significantly reduced.

To determine the correct diagnosis and administer high-quality treatment, it is necessary to identify accurately the physiological processes that establish the flow of blood in the cardiovascular system. Accurate knowledge of the interplay between the blood dynamics and physiological rhythms, such as heart rate, is of crucial importance in treatment applying [25, 26]. In what follows the focus is put on a computational model of the cardiovascular system that represents a resulting human response on such health deficiencies as a myocardial infarction, congestive cardiac failure, chest pain, severe headache requiring hospitalization, as well as other cardiovascular system illnesses, and the use of this model in the analysis of experimental observations from a specific group of people who suffer from transient adaptation after undergoing clinical therapy.

Due to the space limits we cannot and do not require for giving an extensive exposition of such a complex domain as cardiology. The subjects presented in this

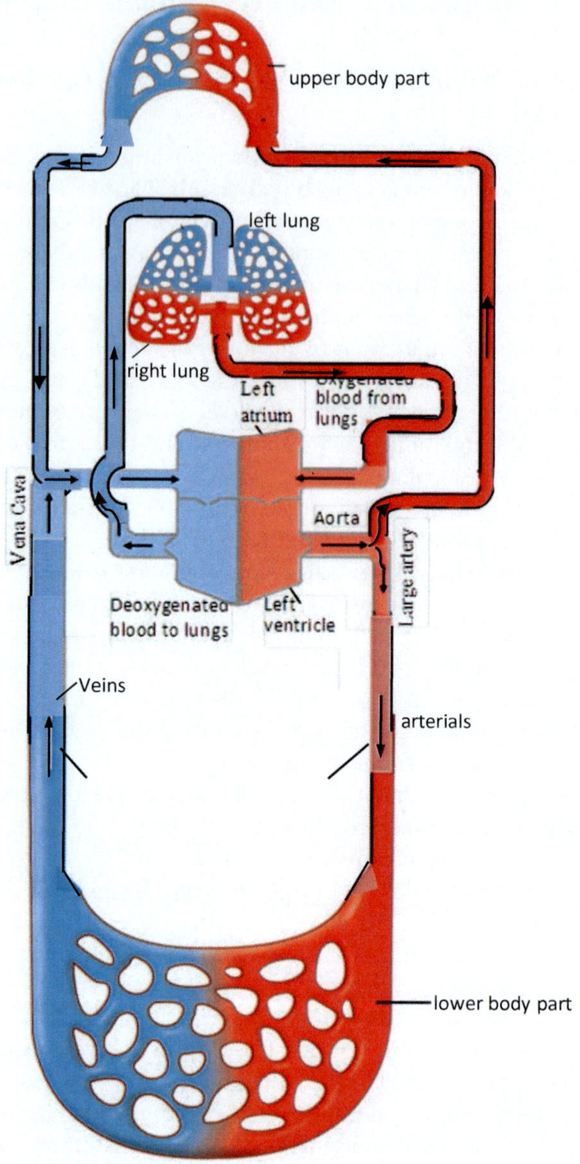

Fig. 3.7 Scheme of a cardiovascular system

section are therefore highly selective and focus on those aspects needed to appreciate the biological background of the dynamical processes generating MMOs. For an in-depth discussion of remaining aspects of cardiology we refer the reader to the corresponding literature positions mentioned at this chapter end.

One of the fundamental role of the heart is to forcing blood venous return by maintenance right atrial pressure low and adopt left ventricle output to the body demand for

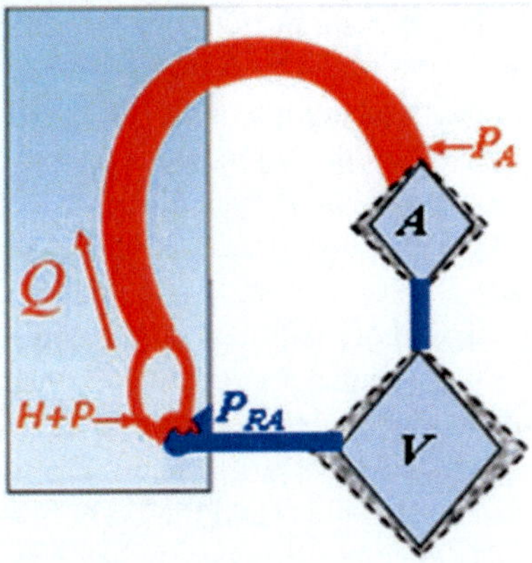

Fig. 3.8. Scheme of systemic circulation in diastole: H + P—hearth and pulmonary, Q—flow, equals cardiac output, A—arterial blood, PA—arterial pressure, V—venous blood, P_{RA}—right atrial pressure

oxygen supply. An improper physiological response to changes in different segments of the cardiovascular system is revealed by variations in blood pressure, which often cause a significant deterioration of the health of the person touched by these changes. It is worth point out that the electrical stimulation that generates the heartbeat resides within the cardiac muscle. The heart beats in the absence of any nervous connections. Many different factors affect and control the cardiac output including the atrial and ventricular reflexes, the autonomic nervous system, hormones, blood ion concentrations, and emotions (Fig. 3.8). These factors affect cardiac output by changing heart rate and stroke volume. After initiation, the electrical signals spread throughout the heart, reaching every cardiac cell rapidly as possible. They go from the top to the bottom of the heart. The actions of all individual cells are strictly coordinated. Through selective openings and closings of plasma membrane channels for sodium, potassium, and calcium ions, the electrical activity of cardiac cells is excited. Through a conductive system, electrical impulses are sent to the chambers that through contractions pump blood into the circulatory system. The most important from point of view of the human body metabolism is the left ventricle with the aorta and large arteries forming a bridge between the heart and the arterioles, serving firstly as a conduit, and second as a shock absorber. Knowledge of instantaneous state of such systems focuses an important attention in clinical practice and research. For this reason, we will concentrate considerations on the left part of the cardiovascular system. Blood flow through the circulation system occurs from areas of high pressure to areas of lower pressure by lowering the downstream left atrial pressure through the actions of the left heart rather than by increasing the upstream pressure. The feed blood back to the heart occurs from the upstream veins and venules. The pressure in this region is determined by the volume of these vessels and by the elastance of their walls. The arterial pulse pressure is created by the single impact volume ejected by the heart and the resistance to its flow from the arterial chamber and by the volume remaining in the aorta at the diastole end. With the same arterial pressure, widely different stroke volumes can be observed, depending on the arterial conductance. The transmission of pressure processes along the arterial circuit and the degree of residual pulsate at the input of peripheries depends on inserted parameters of the arterial circuit. As a consequence, a total peripheral conductance (TPC) changes in complex and nonlinear form depending not only on mean pressure and flow values but also on other dynamical factors such as the shape of the systemic input wave. The stressed component of the whole blood volume determines a blood flow denoted by Q. The flow volume is all the volume that moves through the system on each beat. The blood flow, venous return, and right atrial pressure are determined by the interaction of the function of the heart, and the function of venous return.

The systemic circulation is formed by the arteries, that carry the oxygenated blood pumped by the left ventricle to the living tissues, and the veins that permit the non-oxygenated blood to setback into the right ventricle. The replacement of oxygen between the blood and the body tissues takes place in the microvasculature, which really separates the systemic arterial structure from the venous systems.

The human cardiovascular system (CVS) performs several important functions and the most important from them is carrying the oxygenated blood and other nutrients to the body's organs, particularly to the brain and central nervous system, and to remove carbon dioxide and other harmful products from them. In general, analyzing processes arising in the cardiovascular system is too complex in terms of theory, and therefore, it has become a field of adequate modeling [39].

3.4.2 *Thermodynamic Model of the Cardiovascular System Dynamics*

Cardiovascular systems are composed of many different physiological components (subsystems) and they are effects of their mutual interaction that combine to produce heart rate variability (HRV) as the temporal variation between sequences of consecutive heart beats.

Problems of the nature of blood dynamics, the coupling of the cardiac output motion, and the contribution of peripheral conductance to the blood pressure, respectively, could be analyzed in detail applying physical principles, offering the possibility of generating realistic data on flow, velocity, pressure, and blood volume.

The cardiovascular system can be characterized in terms of its state variables, such as the blood pressure, volume and by the cardiovascular vessel parameters such as compliances and conductances or resistances of the corresponding system's elements. When considering appropriate models of the cardiovascular system, it is very useful in clinical practice to build a model based on physical principles, offering the possibility of generating realistic data on flow velocity, pressure, and blood volume. The cardiovascular system can be modeled using lumped parameters which are hydrodynamic resistance that opposes flow, the compliance which indicates how much volume changes with a given pressure differential, and the inertance which determines the pressure difference required to change the flow rate. Moreover, the laws of physics being applied broadly to the flow of all types of liquids and gases explain also the interaction of pressure, volume, flow, and resistance of the blood in the cardiovascular system.

The blood circulation is accomplished in the arteries, that carry the oxygenated blood pumped by the left ventricle to the living tissues, and in the veins that permit the non-oxygenated blood to setback into the right ventricle. The exchange of oxygen between the blood and the body tissues takes place in the microvasculature, which really separates the systemic arterial structure from the venous systems. In healthy persons, the sinoatrial node, called also the pacemaker of the heart, rhythmically activates excitation pulses 70–80 times per minute, without any neural stimulation. The most important from point of view of the human body metabolism is the left ventricle with the aorta and large arteries forming a bridge between the heart and the arterioles, serving firstly as a conduit, and second as a shock absorber.

A suitably accurate model of the cardiovascular system can be handy in identifying the pressure pulse waveforms produced at the aorta and then in the radial arteries. It can inform greatly about how a system operates in different conditions.

Denoting instantaneous blood flow rate by $q = q(t)$ and the pressure difference by $p = p(t)$, the instantaneous vascular conductance may be expressed by

$$g = g(q, p, t) \tag{3.6}$$

with

$$p = q/g \tag{3.7}$$

what means that the cardiac load, i.e., blood volume per time, coming out of the heart, determines arterial pressure. Here and what it follows the time argument t has been dropped for notational simplicity.

The pumping action of the heart is represented by an impulsive source of the blood flow rate, $q(t)$, that deposits a stroke volume, q_n, into the arterial system during the nth cardiac beat (or cycle in 3.8)

$$q(t) = \sum_n q_n \delta(t - t_n) \tag{3.8}$$

where $\delta(t)$ is a unit Dirac impulse at time $t = 0$ and t_n is the delay time of the nth cardiac cycle. Because the blood is an inert fluid then under the pressure difference between the two ends of a vessel filled with blood the flow cannot change impulsively and the mass of the blood presents the tendency to inertness for moving. The inertness of blood can be modeled by an inertance L, which in respect of arteries relates the pressure drop p with the rate of the blood flow q as follows

$$p = L\frac{dq}{dt} \tag{3.9}$$

Considering the complete systemic arterial tree, we found that the inertial term results from the proper summation of all local inertial terms, and we call it total arterial inertance. It should be emphasized that due to viscous friction in individual parts of the cardiovascular system, kinetic energy of the entire blood flow is transformed into thermal energy due to both dissipative viscosity and turbulent losses. The heat produced by every component of the circulatory system during one heart beat is equal to energy generated by the left ventricle during the systole and is eventually transferred to adjacent tissues. The energy produced by left ventricle is simply

$$E_v = \oint_{\tau} pq dt = \oint_{\tau} p d\theta \tag{3.10}$$

where θ means cardiac blood flow, T is the heart beat period and Γ denotes the closed loop on the (p, θ) plane. Note that the integral is only during systole as this is the only time when ventricular muscles impart new mechanical energy into the blood stream. The dissipation of energy E_v at any component of the circulatory system generally depends on the mean flow rate q, or cardiac output θ, the blood density ρ, the kinematic viscosity μ, the characteristic size of the geometry α, which is governed by the patient body surface area, wall elastic modulus ε, characteristic length l and a dimensionless quantity σ representing the shape of the geometry.

Opposition of a vessel to blood movement is represented by resistance R, which is determined by the radius r and length l of the vessel, and blood viscosity μ, namely

$$R = \left(P_{inp} - P_{out}\right)/\theta \approx 8\,\text{ml}/\pi r^4 \tag{3.11}$$

where P_{inp} and P_{out} denote the mean pressures at the input and output of vessel, respectively. The resistance to flow from the ventricle is therefore mainly found in the resistance vessels: the main aorta and the arteries. When all individual resistances in the ventricle circulation load are properly added, the resistance of the entire segment is obtained and we call this (total) arterial resistance R. It is worth mentioning that heart valves are made of thin leaflets. They do not cause any resistance to the blood during systole and sustain large pressure gradients during diastole. Moreover, unlike vascular vessel wall, they are subjected to very large displacements. These features make the mathematical and numerical modeling of the valve mechanics and of their interaction with the blood flow can be integrated directly to the ventricle activity. As already observed in clinical practice, in cardiovascular system there is mutual exchange of energy between blood and extensible vessel walls. These latter accumulate elastic potential energy under the forces exerted by the blood pressure, which is then transferred to the blood as kinetic energy. From the mechanical point of view, this gives rise to a blood–structure interaction problem. Therefore, apart from the total arterial resistance R, the main part of the arterial load of the heart also includes total arterial compliance C. It is proposed taking full advantage of the information concerning the entire waveforms of pressure and flow, and on this basis estimate the parameter C in accord to the relation

$$q = C\frac{dp}{dt} \tag{3.12}$$

Thus, the rate of change of blood volume in the vessel expressing the cardiac output q is related to a change velocity in time of pressure p inside the vessel with the compliance C. Several studies have shown that compliance C exhibits a strong nonlinear dependence on pressure, which falls sharply from low- to mid-pressure values and tending to an asymptotic value when approaching to high pressure values (like hyperbolic relation) [27].

All the above presented elements, with their nonlinear extensions, can be used in various forms to create models of the heart and its arterial system which together represent relatively simple models of the cardiovascular systems.

Problems of the nature of blood dynamics, the coupling of the cardiac output motion, and the contribution of peripheral conductance to the blood pressure, respectively, could be analyzed in detail applying physical principles, offering the possibility of generating realistic data on flow velocity, pressure, and blood volume. Many aspects of functional interactions that determine the overall behavior of blood flow through the arterial and venous tree can be accurately identified and characterized by its state variables such as blood pressure and volume, as well as cardiovascular parameters such as compliances, inertances, and conductances or resistances of the relevant system components.

Recall, that the so called state variable description provides the system's dynamics as a set of coupled first-order ordinary differential equations in terms of internal variables known as state variables x_i(t), $i = 1, 2, \ldots, n$. Knowledge of those variables at an initial time t_0 and the system inputs (forcing functions) for time $t \geq t_0$, are sufficient to predict the future system state and outputs for all time $t > t_0$ [26, 28, 29]. In general, the choice of state variables is not unique, but usually it is convenient to fixe as state variables those variables which describe the energy stored in all of the independent energy storage elements in the system. However, in the case of cardiovascular system this rule must be slightly modified because apart from the main blood vessels such as arteries (plus smaller arterioles), there are also capillaries and veins (plus smaller veins). The arterial and venous systems are branched in a dendrite-like manner, without any loops. Through their walls they transport the blood without much substance exchange. On the other hand, the "capillary bed", is a dense structure of very thin vessels with perforated walls that allow substance exchange with the intercellular environment. This structure is not tree like, but has many nodes and parallel branches. The capillary structure presents as a porous medium which can transport the blood from any point to other neighboring points, based on the pressure gradient. This porous medium is supplied with blood through a dendrite-like system of arteries (and smaller arterioles), and then the blood is evacuated from it by the venous dendrite-like structure. The topology of the two dendrite-like networks is generated in accord to a minimum cost criterion, coupled with necessary pressure and flow constraints. The systemic capillary structure receives the oxygen-rich blood which is pumped by the heart, and then the oxygen diffuses into the intracellular environment which is very low in oxygen. Moreover, this structure is also the one with the least friction losses. All this energy is eventually transformed irreversibly to heat. Irreversibility found in the blood capillary structure and the arterial vessels are related to energy lost because of dissipation in laminar or turbulent flow. Thus, the heart and the blood circuitry need some energy to work. All this energy is eventually degraded irreversibly to heat. The heart is actually a pump, which first converts chemical energy into mechanical work (through the heart's muscle, the myocardium) and then uses this work to pump blood through the ventricle to the whole circuitry of the cardiovascular system. Due to the large number of parallel branches in the capillary structure, its action is reflected by conductance $g = g(t)$ called peripheral conductance, which is determined by the expression

$$g = G_{min} + \left[1 - \exp\left(-\frac{q^2}{Q_0^2}\right)\right]\frac{pq - Kq^2}{P_C^2} + \exp\left(-\frac{q^2}{Q_0^2}\right)\left(\frac{q^2}{S_m}\right) - \vartheta\frac{dg}{dt} \tag{3.13}$$

where $G_{min}, Q_0, K, P_C, S_M$ are real positive constants, and $\vartheta = \vartheta_0 + \vartheta_1 e^{-\alpha|q|}$, with ϑ_0, ϑ_1, and α being constants, such that $\vartheta_0 \ll \vartheta_1$. When the cardiac output q is small, one can consider $\vartheta \approx \vartheta_1$, while for large cardiac output, $\vartheta \approx \vartheta_0$. Note that in accord to expression (3.13) the peripheral conductance depends not only on q and p but also on the velocity of its changes in time and as such it represents one of the state variables of the model. Thus, expression (3.13) is suitable for fast and large-scale simulation of electrophysiological dynamics of cardiac circuit and is also computationally more tractable for optimization.

Taking into consideration the above findings, we obtain an equation of state variables representing a dynamic cardiovascular system using four first order ordinary differential equations

$$\begin{bmatrix} \dot{x}_1 \\ \dot{x}_2 \\ \dot{x}_3 \\ \dot{x}_4 \end{bmatrix} = \begin{bmatrix} C_p^{-1}(-x_3 + q_s) \\ C_d^{-1}(x_3 - x_2 x_4) \\ L^{-1}(x_1 - Rx_3 - x_2) \\ \vartheta^{-1}(G_{min} + zh + u - x_4) \end{bmatrix} \tag{3.14}$$

where x_1, x_2, x_3, and x_4 are state variables as pressure at input of the aorta, pressure at input of peripheral vessels, blood flow rate in the circuit and conductance of peripheral vessels, respectively; the dot is taken as a symbol for differentiation with respect to time. The components

$$\begin{aligned} z &= 1 - \exp\left(-\frac{x_2^2 x_4^2}{a^2}\right); \\ h &= \frac{x_2^2 x_4 - k x_2^2 x_4^2}{b^2}; \\ u &= \exp\left(-\frac{x_2^2 x_4^2}{c^2}\right) x_2^2 x_4 / d. \end{aligned} \tag{3.15}$$

are strongly nonlinear functions of state variables and threshold parameters of the cardiovascular peripheral elements: *a*, *b*, *c*, d and *k*. The time argument *t* has been omitted for notational simplicity.

In the above model C_p and C_d represent compliances of the highly elastic arteries proximal to the heart and, the less elastic arteries distal from the heart, respectively. The compliance C_p was included to model of ventricle and aorta tissues and to act as a low pass filter whose role is to cutoff the high frequencies of the pulsate rigor of circulation in large vessels, so that the very low pulsation of microcirculation can be reconstructed. A systemic circulation compartment is represented with fixed compliance C_d. The inertance *L* captures the effects of arterial blood pressure pulse propagation on the distal and proximal arterial blood pressure waveforms. The loss of energy in the main vessels of the circulatory system in result of dissipation of a significant part of the energy generated by the ventricle during its compression is mapped by the resistance *R*. The cardiac output is represented by the source $q_s(t) = q_s(t + T)$ with *T* denoting the cycle period. The presented model captures the low-frequency dynamics of the central and peripheral arterial blood pressure waveforms quite well. Actually, it captures the morphology of the aortic blood pressure wavelet quite well, apart from the high-frequency dicrotic notch [19, 21]. An interesting point related to the presented model is that the problems of the nature of blood dynamics, the coupling of the cardiac output motion, and the contribution of peripheral conductance to the blood pressure, respectively, could be with ease analyzed in detail.

Many features of the dynamics of such systems can best be understood in terms of the concepts and laws of thermodynamics. Such formulation of cardiovascular dynamics, which applies to both equilibrium and transient behaviors, can also explain the mathematical and physical foundations of the application of the principles of thermodynamics of nonequilibrium and irreversibility. Focusing for now strictly on the cardiovascular system, we note that the friction between blood and walls of vessels and heart presents some sources of irreversibility. The heart needs some energy to work. All this energy is eventually degraded irreversibly to heat. The “thermodynamic model” represents a mathematical description of the cardiovascular system in which internal states are not exactly known, although statistical information is available about them, and the focus is put on the behavior of the interbranch system, especially on the exchange of energy with the surrounding environment. The terminologies of cardiovascular models and thermodynamics are reflected clearly in the present framework because a distinction between “conservation of energy” and “the first law of thermodynamics” is made. The former term will be used for the sample-path statement of the conservation law, while the term “first law of thermodynamics” will be reserved for the ensemble averaged version. The framework for these considerations encompasses transient behavior, with equilibrium and steady states studied as special cases. Therefore, the first law is stated in terms of the derivatives of various quantities with respect to time.

Recall that a fundamental assumption, proved by experience, of statistical mechanics and thermodynamics is that, if the temperature of a system is strictly positive and its inputs are left undisturbed, the system will eventually settle to a unique state of thermodynamic equilibrium. One reason why the results in this section are not entirely general is that that we have adopted direct access to compliance and inertance to identify relevant state variables. Having such possibilities eliminates the limitation of the basis for explaining the cardiovascular system operation which is closest to the principles of classical thermodynamics. Lifting this limitation is not difficult, because in nonlinear systems the lack of a large signal for thermal noise in elements in which there is a non-linear transformation of potential and kinetic energy into thermal energy is widely accepted.

3.4.3 MMOs as an Indicator of Illness in the Cardiovascular System

Particularly significant value of the established model (3.14) is that that it is capable of accounting for a notable amount of the dynamical behaviors exhibited by the cardiovascular system and especially the identification of the defective functioning of its different segments.

Furthermore, one of many significant advantages of the model is that it gives possibilities for testing theories concerning a particular system. For instance, simulating the blood flow in an aorta, can improve identification of the effects of coupling between the left ventricle and arterial geometry on the flow and, in turn, medical treatment evaluation. This numerical approach is one aspect of a new paradigm in clinical practice known as „predictive medicine" in which the progress in computational fluid dynamics as well as the increase in computer potency has transferred the numerical experiments to the tools being daily in disposition of medical researchers. Thanks to accurate computer simulations there are today possibilities of a detailed identification of the local hemodynamic patterns and of their relations with the vessel instantaneous states of particular segments of the cardiovascular system.

Due to the complex form of the established model (3.14), the determination of its analytical solution is not an easy task and therefore, to examine the influences of diverse system parameters on the hemodynamic process, we can perform computer simulations applying program MATLAB with using effective procedures of numerical integration [25].

Simulations of the blood circulation enable studies of the impact of important parameter values of the cardiovascular model corresponding to the basic physicochemical parameters on the dynamics of blood flow in the circulatory system. The supreme and ambitious purpose of mathematical models applied in medicine is to improve quality of patient life and prevent the health detriment, and then to achieve a right blood pressure at stable normal standard. Therefore, we start by considering the blood flow rate—pressure relations and then turn toward the load reactions and coupling mechanisms responsible for different patterns of the heart output quantities and to identify the relationship between them.

Responses of cardiovascular systems with some pathological changes in the arterial segments are shown in Fig. 3.9. Equations (3.14), (3.15) were solved with the

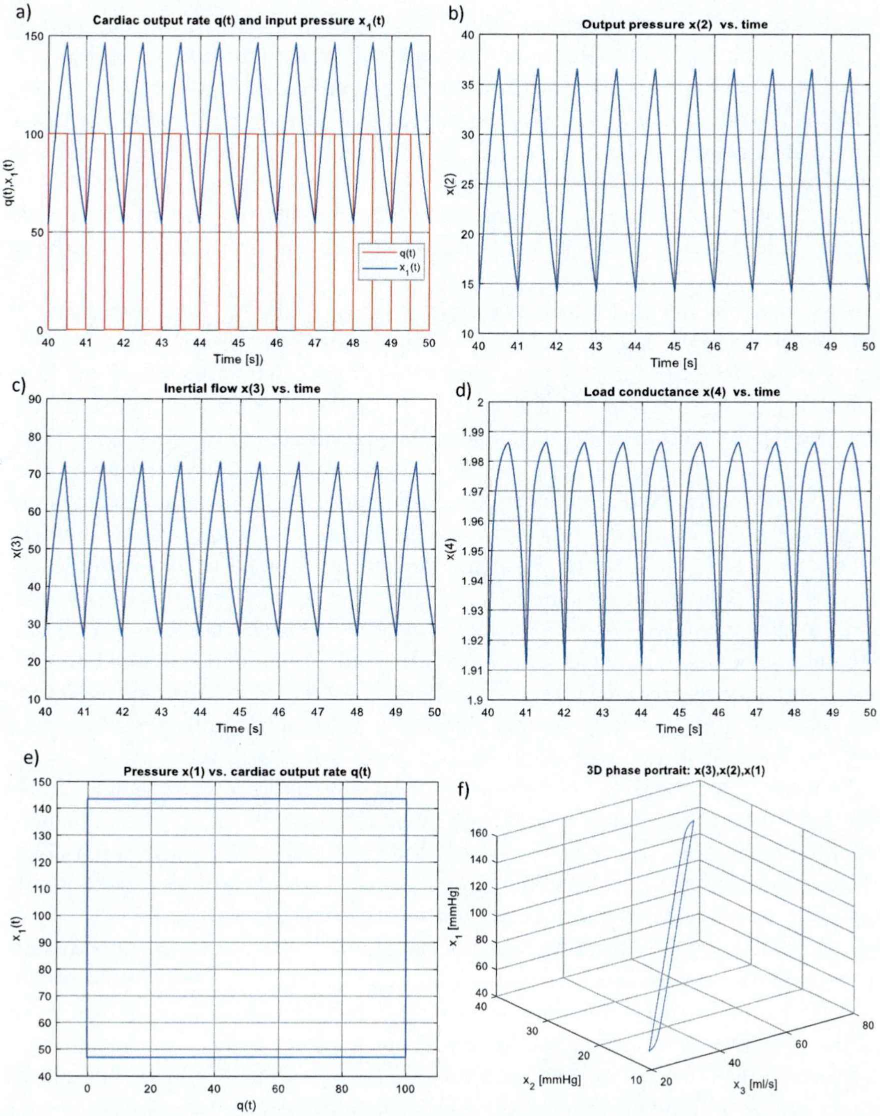

Fig. 3.9 Steady-state time evolution of cardiovascular variables generated by the nominal circulatory model of a heart: **a** cardiac output rate and pressures at input, **b** pressure at output, **c** inertial flow, **d** peripheral vascular conductance, **e** phase portrait $x_1(t)[q(t)]$, f) 3D phase portrait

variable step ODE23 procedure from MATLAB with $RelEr = AbsEr = 10^{-8}$ and $0 \leq t \leq 50$ s for different initial conditions. Figure 3.9 illustrates the steady-state pulsatile arterial blood pressure waveforms that result from simulating the model (3.14), (3.15) with $Q = 100$ mL/s, $T = T_n = 1$ s, $R = 1$ mmHg/(mL/s), $L = 0.001$ mmHg/(mL/s2), $C_p = 0.25$ mL/mmHg, and $C_d = 0.025$ mL/mmHg. These parameters represent typical values for a 70 kg male human [7, 26], and, when used with Eqs. (3.14), (3.15), result in reasonable approximations of the pressure waveforms during the cardiac cycle. Periodic orbits reflect the deterministic dynamics of the underlying system.

It is worth to notice that the generated waveforms of the blood pressures, inertial flow, and peripheral conductance change their values with the same period as the heartbeat and take the form of relaxation oscillations. The slopes of the pressure curves (Fig. 3.9a) in the aorta and at the exit of the arteries (Fig. 3.9b) are the same, but during their descents the slope of the pressure curve for the output of the arteries is steeper than at the aorta entrance. At the same time, the pressure drop in the systemic arterial system between the ascending of the aorta and the equivalent artery does not exceed 105–110 mmHg at the upper level and 40.5–41.5 mmHg at the lower level. The 2D and 3D phase portrait are formed by closed univalent curves which correspond very well to the relaxation oscillations. It is also visible that the instantaneous vascular conductance (Fig. 3.9d) changes its values in a slightly different manner and exhibits the form like the graph of rectified sine function with small magnitude superposed on relatively large constant value. The dependence of aortic pressure on blood flow rate (Fig. 3.9e) takes the typical human left ventricular pressure–flow rate loop.

Particularly dignified of attention is that that the cardiovascular model (3.14) with (3.15) is capable of accounting for a notable amount of the dynamical behaviors exhibited by the cardiovascular system and particularly the identification of the defective functioning of its different segments. Responses of cardiovascular systems with some pathological changes in the arterial segments are shown in Fig. 3.10. In this case, the steady-state response waveforms exhibit the form of mixed-mode oscillations (MMOs) caused by changes in compliance values ($C_\mathrm{p} = 0.025$ mL/mmHg, $C_\mathrm{d} = 0.025$ mL/mmHg), inertance ($L = 0.0001$ mmHg/(mL/s^2)), resistance ($R = 0.0001$ mmHg/(mL/s)), magnitude of blood rate pulses ($Q = 70$ mL/s) and only slightly modified peripheral conductance. These waveforms express the role of the arterial flexibilities in shaping the MMO patterns (generation of STOs—sub-threshold oscillations—and the onset of spikes) and the blood dynamic mechanisms that give rise to MMOs. Along with the change of the parameter of the element affected by the pathological change, MMOs may appear alternating between SAO and LAO [4, 5]. The MMO waveforms in Fig. 3.10a can be denoted as 3^3 MMOs. It is worth to mention that the generation of MMO patterns requires the coordinated action of such processes as: (*i*) generation of SAOs, (*ii*) a spiking, including the onset of spikes (LAOs) and a support of the spiking dynamics, and (*iii*) a return phenomenon from the spiking regime back to the sub-threshold regime. Moreover, generation of MMO patterns requires an additional mechanism that provokes the dynamic transition between these two regimes by an oscillatory forcing input (cardiac output) and

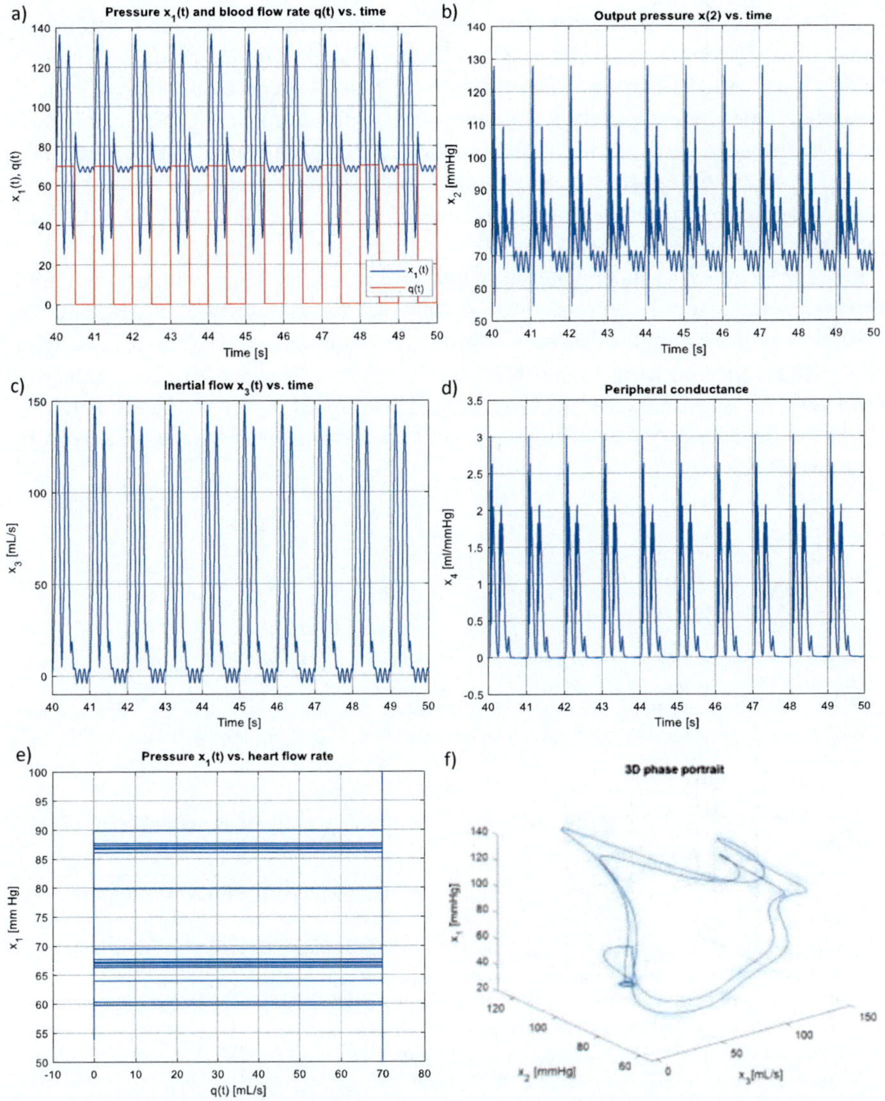

Fig. 3.10 Steady-state time evolution of cardiovascular variables generated by a pathological circulatory model of a heart: **a** pulsatile left ventricle blood output rate and MMOs pressure at aorta input, **b** pressure MMOs at arterial output, **c** MMO oscillations of inertial flow rate, **d** MMOs of peripheral conductance, **e** output pressure versus blood flow rate at ventricle output, **f** 3D phase portrait of MMOs

by an additional nonlinear dependent variable (peripheral conductance) leading to the 3D canard phenomenon. The major problem related to mixed mode oscillations (MMOs), which has not been studied so far in relation to cardiovascular systems, is focused on determining the correlation between different forms of response identified by the model with the states of pathological changes in the vascular elements of a given patient. In some cases of nonlinear dynamics, the effect can be fairly complex, with a sequence of mixed-mode oscillations of different standing and numbers of SAOs and LAOs.

In some cases of weakened action of cardiovascular elements with an increased non-linear effect, more complex disease states of the patient with a mixed mode oscillation sequence with different positions and numbers of SAO and LAO may occur. Therefore, the form of the MMOs may exhibit the degree of nonlinearity produced by a disease of the cardiovascular system and diagnosis of the patient's instantaneous state could be checked out via a function of heartbeats in accord to LAOs and SAOs.

References

1. Izhikevich, E.M.: Dynamical Systems in Neuroscience: The Geometry of Excitability and Bursting. MIT Press, Cambridge (2007)
2. Lu, Z., Mukkamala, R.: Continuous cardiac output monitoring in humans by invasive and non invasive peripheral blood pressure waveform analysis. J. Appl. Physiol. **101**(2), 598–608 (2006)
3. Mukkamala, R., Reisner, A., Hojman, H., Mark, R., Cohen, R.: Continuous cardiac output monitoring by peripheral blood pressure waveform analysis. IEEE Trans. Biomed. Eng. **53**(3), 459–467 (2006)
4. Heldt, T., Chang, J.L., Chen, J.J.S., Verghese, G.C., Mark, R.G.: Cycle-averaged dynamics of a periodically-driven, closed loop circulation model. Control Eng. Pract. **13**(9), 1163–1171 (2005)
5. Parlikar, T.A., Heldt, T., Verghese, G.C.: Cycle-averaged models of cardiovascular dynamics. IEEE Trans. Circuits Syst. I Fundam. Theory Appl. **53**(11), 2459–2468 (2006)
6. Shibao, C., Grijalva, C.G., Raj, S.R., Biaggioni, I., Griffin, M.R.: Orthostatic hypotension-related hospitalizations in the United States. Am. J. Med. **120**, 975–980 (2007)
7. Fedorowski, A., Stavenow, L., Hedblad, B., Berglund, G., Nilsson, P.M., Melander, O.: Orthostatic hypotension predicts all-cause mortality and coronary events in middle-aged individuals. Eur. Heart J. **31**, 85–91 (2010)
8. Kozlovskaya, I.B., Grigoriev, A.I.: Russian system of countermeasures on board of the International Space Station (ISS): the first results. Acta Astronaut. **55**, 233–237 (2004)
9. Walchenbach, R., Geiger, E., Thomeer, R.T., Vanneste, J.A.: The value of temporary external lumbar CSF drainage in predicting the outcome of shunting on normal pressure hydrocephalus. J. Neurol. Neurosurg. Psychiatry **72**, 503–536 (2002)
10. Arnold, A.C., Raj, S.R.: Orthostatic hypotension: a practical approach to investigation and management. Can. J. Cardiol. **33**, 1725–1728 (2017)
11. Low, Ph.A., Tomalia, V.A.: Orthostatic hypotension: mechanisms, causes, management. J Clin Neurol **11**(3), 220–226 (2015). https://doi.org/10.3988/jcn.2015.11.3.220
12. Adamec, I., Junakovic, A., Krbot, M., Habek, M.: Association of autonomic nervous system abnormalities on head-up tilt table test with joint hypermobility. Eur. Neurol. **79**(5–6), 319–324 (2018)

13. Genecand, L., Dupuis-Lozeron, E., Adler, D., Lador, F.: Determination of cardiac output: a game of thrones. Respiration **96**(6), 1 (2018)
14. MacEwen, C., Sutherland, Sh, Pugh, D.J., Tarassenko, L.: Validation of model flow estimates of cardiac output in hemodialysis patients. Ther. Apher. Dial. Off. Peer-Rev. J. Int. Soc. Apher. Jpn. Soc. Apher. Jpn. Soc. Dial. Therapy **22**(4), 1 (2018)
15. Niewiadomski, W., Gasiorowska, A., Krauss, B., Mróz, A., Cybulski, G.: Suppression of heart rate variability after supramaximal exertion. Clin. Physiol. Funct. Imaging **27**(5), 309–319 (2007)
16. Strasz, A., Niewiadomski, W., Skupi ́nska, M., Gąsiorowska, A., Laskowska, D., Leonarcik, R., Cybulski, G.: Systolic time intervals detection and analysis of polyphysiographic signals. In: Jablonski, R., Brezina, T. (eds) Mechatronics Recent Technological and Scientific Advances. Springer, Berlin, (2011)
17. Rajzer, M., Kawecka-Jaszcz, K.: Arterial compliance in arterial hypertension. From pathophysiology to clinical relevance. Arter. Hypertens. **6**(1), 61–73 (2002)
18. Dimitrakopoulos, E.G.: Nonsmooth analysis of the impact between successive skew bridge-segments. Nonlinear Dyn. **74**, 911–928 (2013)
19. Rotstein, H.G., Oppermann, T., White, J.A., Kopell, N.: A reduced model for medial entorhinal cortex stellate cells: subthreshold oscillations, spiking and synchronization. J. Comput. Neurosci. **21**, 271–292 (2006)
20. Acker, C.D., Kopell, N., White, J.A.: Synchronization of strongly coupled excitatory neurons: relating network behavior to biophysics. J. Comput. Neurosci. **15**, 71–90 (2003)
21. Krupa, M., Szmolyan, P.: Relaxation oscillation and canard explosion. J. Differ. Eqs. **174**, 312–368 (2001)
22. Hodgkin, A., Huxley, A.: A quantitative description of membrane current and its application to conduction and excitation in nerve. J. Physiol. **117**, 500–544 (1952)
23. Zhang, X.J., You, G.Q., Chen, T.P., Feng, J.F.: Readout of spike waves in a microcolumn. Neural Comput. **21**, 3079–3105 (2009)
24. Nayak, S.K., Bit, A., Dey, A., Mohapatra, B., Pal, K.: A review on the nonlinear dynamical system analysis of electrocardiogram signal. Hindawi J. Healthc. Eng. **69**, 1–19 (2018).
25. Heldt, T., Verghese, G.C., Mark, R.G.: Mathematical modeling of physiological systems. In: Batzel, J.J., Bachar, M., Kappel, F. (eds) Included in Mathematical Modeling and Validation in Physiology: Applications to the Cardiovascular and Respiratory Systems. Springer, Berlin (2013)
26. Subramaniam, B., Khabbaz, K.R., Heldt, T., Lerner, A.B., Mittleman, M.A., Davis, R.B., Goldberger, A.L., Costa, M.D.: Blood pressure variability: can nonlinear dynamics enhance risk assessment during cardiovascular surgery. J. Cardiothorac. Vasc. Anesth. **28**(2), 392–397 (2014)
27. Tharp, B.Z., Erdmann, D.B., Matyas, M.L., McNeel, R.L., Moreno, N.P.: The Science of the Heart and Circulation. Baylor College of Medicine, Houston (2009)
28. Solinski, M., Gieraltowski, J.: Different Calculations of Mathematical Changes of the Heart Rhythm. Center of Mathematics Applications, Warsaw University of Technology, Warsaw (2015)
29. Giusti, A., Mainardi, F.: A dynamic viscoelastic analogy for fluid-filled elastic tubes. Meccanica **51**(10), 2321–2330 (2016)
30. Trzaska, Z.: Nonsmooth analysis of the pulse pressured infusion fluid flow. Nonlinear Dyn. **78**, 525–540 (2014)
31. Trzaska, Z.: Dynamical processes in sequential-bipolar pulse sources supplying nonlinear loads. Electr. Rev. **90**(3), 147–152
32. Trzaska, Z.: Properties and applications of memristors—memristor circuits with innovation in electronics. In: Czy˙Z, Z., Macia, G, K. (eds) Contemporary Problems of Electrical Engineering and Development and Evaluation of Technological Processes, pp. 76–93. Publisher TYGIEL, Lublin (2017)
33. Lesniak, M.S., Clatterbuck, R.E., Rigamonti, D., Williams, M.A.: Low pressure hydrocephalus and ventriculomegaly: hysteresis, non-linear dynamics, and the benefits of CSF diversion. Br. J. Neurosurg. **16**, 555–561 (2002)

34. Rowell, L.B.: The cardiovascular System. In: Tipton, C.M. (ed) Exercise Physiology. American Physiological Society, Rockville (2003)
35. Colombaro, I., Giusti, A., Mainardi, F.: On the propagation of transient waves in a viscoelastic Bessel medium. Z. Angew. Math. Phys. **68**(3), 62 (2017)
36. Marszalek, W.: Bifurcations and Newtonian properties of Chua's circuits with memristors. DeVry Univ. J. Sch. Res. **2**(2), 13–21 (2015)
37. Marszalek, W.: Fold points and singularity induced bifurcation in inviscid transonic flow. Phys. Lett. A **376**, 2032–2037 (2012)
38. Podhaisky, H., Marszalek, W.: Bifurcations and synchronization of singularly perturbed oscillators: an application case study. Nonlinear Dyn. **69**, 949–959 (2012)
39. Wolf, A., Swift, J.B., Swinney, H.L., Vastano, J.A.: Determining Lyapunov exponent from a time series. Physica **16D**, 285–317 (1985)

Chapter 4
MMOs in Chemistry

4.1 Preliminary

In a present competitive world of globalization, activities of chemical industries specialists are striving to improve their processes and products for the dynamic and high expectations from consumers. However, the chemical businesses can move their attention to the cost challenges if they are able to fabricate innovative products that will carry a higher value to the customers, even if the cost of final products is higher. The most common challenges in almost all chemical engineering fields are manufacturing high quality products with the lowest cost and generating abundant profit from market sales. Successful competitive chemical branches are not only able to improve their product costs, but also capable to bring a high value profit through their product innovations. The innovations in this field are becoming more challenging in developing products that have different applications. For this reason, chemical engineers need a systematic and reliable methodology to innovate their products, especially through current problem solving and forecasting. Therefore, it is critical for the investigations in chemical engineering to have a strong and reliable approach and method to support the product innovation and development. The challenge consists of problems that have various levels of complexity, constraints and limitations that represent various levels of difficulties. Problem-solving approach in chemical engineering requires the use of numerical language and software. However, the problem-solving scale only covers on optimization such as to generate the model equations for the problem at hand, appoint an appropriate numerical method to solve the model, write and debug a program to solve the problem using the fitted numerical algorithm, analyze and estimation of the obtained results for validity and precision. In the current condition, investigations in the chemical engineering show a higher constraint in innovation and creativity compared to research in other engineering fields. In this context, it is critical to have effective methodologies and tools by chemical researchers to develop interest in methods development leading to innovation for the competing applications. The main reason is that a typical sequence of events may be such that smooth changes of a control parameter even in a relatively

Z. Trzaska, *Mixed Mode Oscillations (MMOs)*, Studies in Systems, Decision and Control 374, https://doi.org/10.1007/978-3-030-76867-6_4

simple chemical dynamical system, described by deterministic nonlinear differential equations, can cause a sudden transition to a completely new, qualitatively different state, what means establishing a non-equilibrium stationary state. When a control parameter such as temperature, flow rate, or influx concentration is under variation, bifurcation and transition of the system from one dynamic state to another occurs. Periodic or nearly periodic temporal changes of concentrations in the reacting system provide a systematic impulse to improve knowledge and develop new findings in the field of non-linear chemical dynamics. Oscillating chemical reactions can be made extremely sensitive to some components in result of modifying the system structure or determination in the non-equilibrium stationary state bifurcation, providing necessity of applying new methods for their investigations. When the differential equations, which present the model of a given system, are nonlinear, their analytical solution is often unknown and one has to use the computer to solve the problem numerically. In track of computer calculations not only numerical solutions of various problems are obtained, but also completely unexpected phenomena can be discovered [17]. It should be emphasized that mixed-mode oscillations, for which the oscillatory cycle consists of a number of large excursions combined with a number of small peaks, are frequently observed in both experiments and models appropriate to chemical and biological systems [1–11]. They have significant applications in nonlinear chemistry and medicinal chemistry. The progresses in the description of systems combining nonlinear chemical kinetics and the mixed mode oscillations are presented in the sequel.

4.2 Oscillatory Chemical Systems

Chemical oscillations are the result of competing kinetic processes involving nonlinear feedbacks. In chemical reactions there exists some equilibrium mixture of reactants and products. Practically, the instantaneous reaction of aqueous solutions of silver nitrate and sodium chloride is a good example. On the other hand, the most widely known oscillatory chemical reaction is certainly the Belousov–Zhabotinsky reaction [18]. The oscillations occur in the concentrations of various intermediate species, including Br^-, $HBrO_2$, and the redox catalyst. For most of this stage of the reaction, the amplitude and period of individual oscillations varies only slightly from that of the previous state. The reaction typically shows a preoscillatory initial phase, followed by an oscillatory phase that may run for several hours in a closed reactor. However, the oscillatory proceeding is strictly transient and the effects of reactant expenditure cause an end of oscillation and a final monotonic entrance to the chemical equilibrium state. Note that periodic or oscillating reactions may occur even in simple reaction systems and result from a series of consecutive autocatalytic reactions.

Because chemical reactions have no momentum they are unlike mechanical oscillators such as a mass on a spring or the head of a drum. However, in a chemical

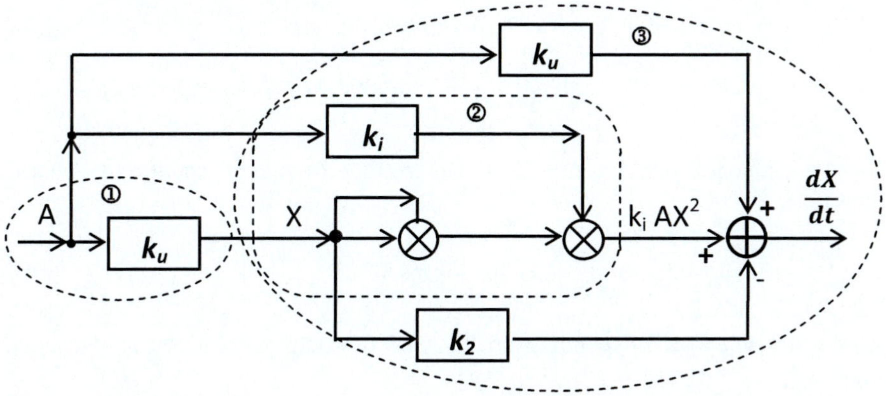

Fig. 4.1 Block scheme illustration of a chemical oscillatory mechanism: ①—initiation, ②—autocatalyst, ③—scavenging

oscillator, the values of certain parameters change cyclically in time around the state of equilibrium.

A simple formal example of an oscillatory mechanism including positive and negative feedback steps is the model presented by block scheme in Fig. 4.1.

The rate equation for X is given by

$$\frac{dX}{dt} = k_i AX^2 + k_u A - k_2 X \tag{4.1}$$

The consumption of concentrate A in reaction (4.1) also limits the rate of increase of product X and can be considered as a negative feedback. However, this contribution to the rate term, directly linked to the production of X, cannot lead to an oscillatory instability in an homogenous system. In the absence of the autocatalytic term in reaction (4.1), this would be a simple linear cascade of reactions that could not produce oscillations. Note, that the autocatalytic reaction (4.1) exerts a positive feedback on component X while reaction $-k_2$X which independently opposes to the increase of X is a negative feedback process. Positive feedbacks are always destabilizing processes, while stabilizing ones are the antagonist negative feedbacks. Anything stating on the complexity of the kinetic mechanism, oscillations develop only when the positive feedback evolves on a shorter time scale τ_p than the negative feedback τ_n. In a strict sense, there must be a "delay" between the positive and negative feedbacks actions in the given system.

Generally, in the chemical world the first principal source of nonlinearities leading to instabilities is involved by the local reactive dissipative processes. Thus they may be exhibited even in the absence of spatial degrees of freedom, as for instance in a well-mixed reactor, what is leading to a wealth of nonlinear behaviors. This places nonlinear kinetics in chemical systems at the forefront for understanding the origin of endogenous oscillating and model phenomena [21].

In real chemical systems, autocatalysis usually occurs in a multistep process, like

$$A + nX \rightarrow mY \quad \text{with}\, m > n \tag{4.2}$$

$$Y \rightarrow \cdots \rightarrow X \tag{4.3}$$

where the sequence of reactions (4.3) is fast, so that the rate of reaction (4.2) is as follows

$$\frac{dX}{dt} = (m - n)k_2 A^{\alpha} X^{\beta} \tag{4.4}$$

with constants α and β depending on the kind of chemical reaction and reactor structure.

Usually, chemical reactors can present an open form or a closed one. In closed systems, all chemical reactions spontaneously evolve to their thermodynamic equilibrium, and in such conditions, it results that oscillations may only be observed for a brief interval of time, when the chemical composition is still at a finite difference of quantities with respect to an equilibrium state [22, 24]. Actually, batch oscillatory reactions are only very few. They require that the initial great quantity of reagents be sparingly utilized at each oscillatory cycle.

The simplest open chemical reactor is the so called Continuous Stirred Tank Reactor (CSTR). It is a mixture of chemicals in a fixed volume tank, stirred energetically and permanently refreshed by constant inflows of reagents (Fig. 4.2). In the ideal case, the input flows are supposed to instantaneously and uniformly mixing into

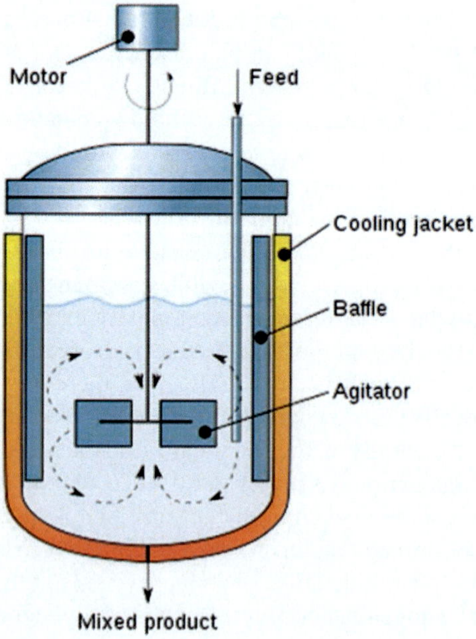

Fig. 4.2 An isothermal continuous stirred tank reactor

the contents of the tank with a conservation of the input volume at any time. In what follows the liquid phases of chemistry will be considered although the methodology can be applied to a wider class of systems. It is worth noticing that the CSTR is playing a crucial role in the investigations in the field of nonlinear chemistry when the acting between reaction and diffusion is not involved (i.e. well mixed reactor).

It should be emphasized that in the case of nonlinear chemical systems, even finely dispersed residual feed mixing inhomogeneities in the reacting solution can dramatically affect the overall dynamics [25]. In the case of the well mixed reactor, all spatial concentration gradients do not exist. Then the dynamics of reactants is driven only by the kinetic force and input–output balance conditions. Hence, the control and design of new oscillators should take into account the appropriate application of a constant feed of the stirred tank reactor (CSTR). Normally, the oscillatory behavior of a CSTR, is manifested through periodic changes with respect to time in the reaction temperature and/or the concentrations of the reacting species. In order to model such behavior, let us consider a chemical system involving species X and Y with respective concentrations x and y, and ruled by the following reaction mechanisms

$$\begin{aligned} &\mathrm{A} \rightarrow \mathrm{X} \\ &\mathrm{B} + \mathrm{X} \rightarrow \mathrm{Y} \\ &2\mathrm{X} + \mathrm{Y} \rightarrow 3\mathrm{X} \\ &\mathrm{X} \rightarrow \mathrm{Q} \end{aligned} \tag{4.5}$$

where A, B are the initial reactants, Q the final product and X, Y the intermediate products. The initial reactants considered in large excess fulfill the conditions for the CSTR. Thus the system may be considered as open to the external environment and the corresponding kinetics takes the form

$$\begin{aligned} dX/dt &= AX - BY - \alpha X^2 Y \\ dY/dt &= BX + AY - \alpha X Y^2 \end{aligned} \tag{4.6}$$

where X and Y are an activator and inhibitor, respectively. From (4.6) it is easily seen that in the chemical world the first principal source of nonlinearities leading to instability arises from the local reactive dissipative processes. However, two caveats are to be explained. First, from an ideal point of view, it is fair to admit that this system cannot represent a real chemical system in the accurate sense as the "concentrations" X and Y may become negative what is unacceptable. However allowing exact solutions enable the introduction, in not too formal a way, of some important concepts which explain in satisfactory means most of the natural chemical reactions. The other caveat pertains to the coefficients A and B, which can be thought of as representing the concentrations of other species that evolve on a much longer time scale and that may be considered as constant on the timescale considered in the given reaction. In this pool chemical approximation Eq. (4.2) can be considered as representing an open system, the parameters of which may be varied to control the difference with respect to some reference state. Parameter α measures the strength of the nonlinear coupling that introduces the cubic term in an expression describing the concentrations.

From Eq. (4.6) it follows that $X_0 = 0$ and $Y_0 = 0$ is the steady state solution regardless of the values of the parameters. By linearizing around (X_0, Y_0) it is possible to test its stability. The characteristic equation from the determinant of the Jacobian matrix takes the form

$$\det\begin{bmatrix} \lambda - A & B \\ -B & \lambda - A \end{bmatrix} = \lambda^2 - 2\lambda A + A^2 + B^2 = 0 \tag{4.7}$$

Thus the complex conjugate pair $\lambda_{1/2} = A \pm iB$ form the generalized eigenfrequencies of the system. As long as A remains negative ($A < 0$) the state (X_0, Y_0) stays asymptotically stable. When A becomes positive, then the state of the system is unstable and for $t \to \infty$ all trajectories in the (X, Y) plane tend to the circle of radius $R_s = \sqrt{\frac{A}{\alpha}}$. In this case the new state $X(t) = R_s \cos Bt$ and $Y(t) = -R_s \sin Bt$ is stable and the concentrations evolve periodically in time. This corresponds to a limit cycle that coexists with the unstable state $A = B = 0$. The radius of the limit cycle increases with A.

This may be represented by the bifurcation diagram shown in Fig. 4.3. When the real part of the least stable single pair of complex conjugate eigenvalues changes sign, we say that the system undergoes a Hopf bifurcation [26]. The bifurcation is supercritical when the limit cycle is born with an increase of the control parameter above its marginal value $A = 0$.

In a case, when the bifurcation parameter increases and the unstable limit cycle collapses into the steady state we have a subcritical Hopf bifurcation. From the point of view of stability, it shows a limitation of the linear stability analysis.

Often, to stabilize the system, another bifurcation scheme with A_c can be considered (Fig. 4.4), which shows the essential elements of such a process. In the subcritical region, when $A_c < A < 0$, in addition to steady state $X_0 = Y_0 = 0$, there are two limit cycles with different amplitudes. The one with the smallest radius is unstable, and the one with the larger radius, which is stable, keeps the system stable (even in supercritical region). These two limit cycles of different amplitudes provide the basis for the

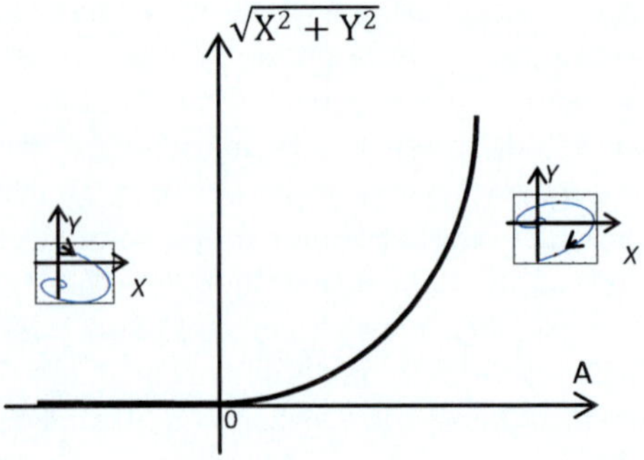

Fig. 4.3 Subcritical Hopf bifurcations for the λ–ω system

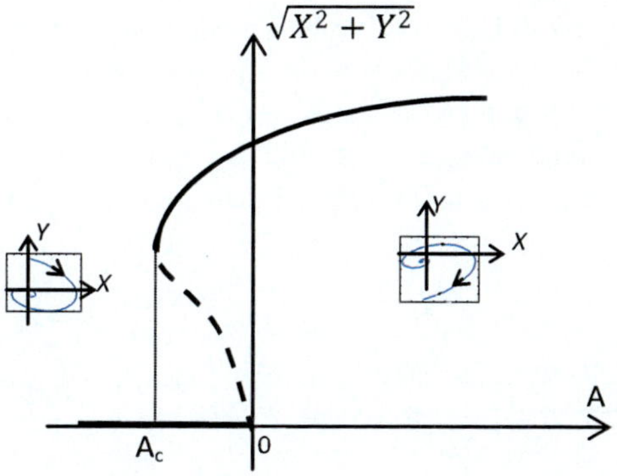

Fig. 4.4 Supercritical Hopf bifurcation stabilized through the interaction with a saddle-node bifurcation

generation of mixed mode oscillations (MMOs). For A_c, the radii of the two cycles become equal and fall within the so-called saddle-node bifurcation, connected by their unstable states.

It is worth noticing that in the vicinity of a supercritical Hopf bifurcation point it may be shown that the dynamics obeys the generic form that is the well known as the complex Stuart Landau equation [12] which constitutes a universal model for oscillators with a weak nonlinearity. Note, that in the phase space spanned by all concentrations the limit cycle is not a circle anymore, although still it forms a closed curve. Moreover, it changes in time with a speed that now depends on the amplitude of oscillations as the equations for amplitude and phase angle can't longer be decoupled. In result, the oscillations lose their harmonic character [19, 20, 23].

4.3 Numerical Simulations

Numerical simulation is a powerful tool in understanding complex solution structure of autocatalytic reaction problem in systems like (4.6) involving two chemical species, a reactant A and an autocatalyst B species. Currently, this is met with considerable attention in everyday chemical practice, which is likely to persist over a somewhat longer period, because it allows the use of computer models to identify deficiencies of processes realized in practice. Thanks to accurate computer simulations, there are today possibilities of a detailed identification of the local chemical reaction patterns and of their relations with the vessel instantaneous states of particular reactants in the system. Varying several parameters of the problem leads to obtain bifurcation surfaces in the parametric space that separate parametric regions differing by the number and nature of states and/or their stability. Any change of the parameters of the system gives the bifurcation when the corresponding trajectory in the parametric space intersects a bifurcation surface. In the sequel the main attention is focused on conditions leading to generation of MMOs with organizing the results obtained by straightforward numerical integration.

To illustrate this phenomenon we will considered a continuously stirred tank reactor (CSTR) where the Iodate-Arsenous Acid (IAA) reaction occurs. The kinetics term of the IAA reaction under the conditions presented above may be written, with a trivial change of variable as $f(x) = ax + bx^3 + c$. If a kinetic coupling with another substance y takes a place, then such system introduces the following description

$$\begin{aligned}\frac{dx}{dt} &= ax + bx^3 + c - y \\ \frac{dy}{dt} &= d(x - gy)\end{aligned} \tag{4.8}$$

where constants a, b, c, d and g determine properties of reacting substances and reaction conditions. Here $x = x(t)$ and $y = y(t)$ represent the dynamical variables in the system. To identify the behavior of the IAA reaction the situation where arsenic acid is in stoichiometric excess and when the pH is buffered forms the base for detailed simulations.

Solutions of this equation are easy to investigate by numerical approach. We have shown above that there is a subcritical one Hopf branch at $c > 0$ in this model; see Fig. 4.5. On the other hand if the control variable c takes a time-dependent "adaptation term" C_{ad} to the external drive c, governed by equations of the form

$$\mathrm{C_{ad}}(t + 0) = C_{ad}(t - 0) - \gamma \tag{4.9}$$

with $\gamma > 0$ and on the order of tenths then the unstable limit cycle is appearing which grows as c decreases and suddenly "explodes," colliding with and destroying the stable limit cycle. Thus, similarly to supercritical canards, subcritical ones can give rise to mixed-mode oscillations (Fig. 4.5). An open system exhibiting kinetic bistability is active at a particular pumping rate in either of two stationary states, and concentrations of the intermediate components may differ by orders of magnitude in each of the states. Kinetic bistability of reactions is of practical importance in

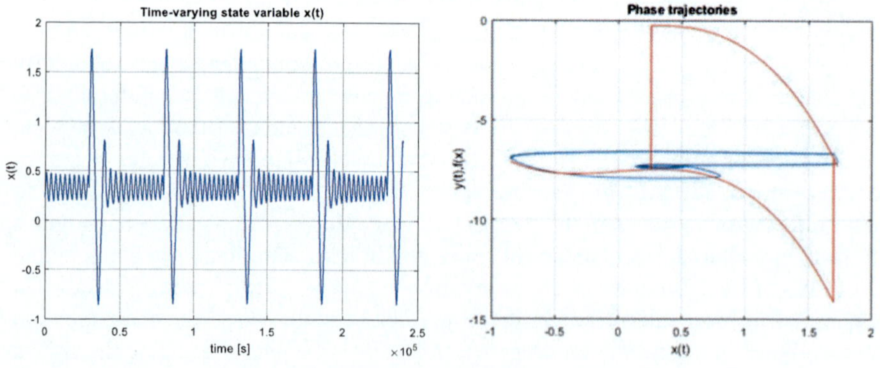

Fig. 4.5 Simulation results of an electrochemical reactor: **a** 1^{13} MMOs, **b** phase trajectory

chemical production processes which utilize pumped tank reactors, and may also be important in biological systems as an on/off mechanism. It is worth recalling that chemical reactions constitute the base in which biological oscillation originates, and the most numerous and important chemical oscillators are found in living systems [18, 28]. Many researchers study chemical systems by investigating the effects of a pharmaceutical on a chemical oscillator and examine the reaction mechanism. Different product concentrations in one stationary state versus another may be of appreciable practical importance in chemical manufacturing processes utilizing pumped tank reactors.

4.4 The Mathematical Model of Electrochemical Reactors

Electrochemical processes have received recently increasing attention in rapidly growing fields of science and technology, mainly in such domains as nanosciences (nanoelectrochemistry) and life sciences (organic and biological electrochemistry). This has a bearing mainly to such industries as the aerospace, automotive, military, microelectronics, computers, energy, biotechnology, medicine and related others. During the last twenty years, electrochemical processes have become the basis for techniques comprehensively being relevant for the production of nanostructured solid materials and surface layers, which are widely applied in many industries that use the top technologies. Presently, challenges in relations to electrochemical processes used in the practice are becoming more and more expressive in terms of better adapting them to the special technological requirements with meeting the directives for reducing negative impact on the surrounding environment. The electrocrystallization process is exceptionally suitable for manufacturing of specific materials, mainly the protective layers, and in the near future, it may be more effective than previously involved approaches in technologies of surface modification of known materials, as well as newly produced ones [27–30].

Taking into account the specificity of electrochemically produced materials, it is possible the formation of such structures of materials and their properties that cannot be obtained by other technologies. The possibility of conscious control of the structures in the nanometric scale leads up to producing the useful materials with new properties and overcome previously insurmountable barriers to the development of technology. Nanomaterials possess at least one dimension sized from 1 to 100 nm [4, 6] and not only unique geometric, mechanical, electronic and chemical properties, but also properties different from macroscopic materials, such as quantum effect, surface effect, small size effect, etc. These properties have greatly prompted a broad range of applications of nanomaterials in medicine, electronics, biotechnologies, environmental science, energy production and biosensors.

All chemical processes in which there is a net change in atomic charge occur as the result of oxidation–reduction reactions also known as redox reactions. They can be represented by a diagram shown in Fig. 4.6.

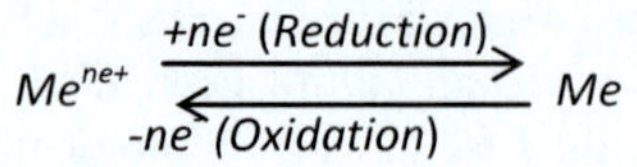

Fig. 4.6 Scheme for redox reaction

Electrocrystallization processes generated as a result of oxidation (losing electrons) and reduction (gains electrons) reactions occur at the electrodes: anode and cathode, respectively. The processes of oxidation and reduction occur simultaneously and cannot happen independently of one another.

Overall, the manufacturing of nanostructured materials is too complex for a correct theory, so it becomes the subject of computer simulations. The most frequently investigated electrocrystallization process is cathodic metal deposition on foreign and native substrates from electrolytes containing simple and/or complex metal ions [3, 12]. Typical examples are the electrocrystallization of Ag from electrolytes containing Ag^+ [13], the cathodic deposition of Ag on n-Si from electrolytes containing $[Ag(CN)_2]^-$ ions [3], and the electrodeposition of Cu [4], from electrolyte containing $[Cu_2SO_4]^+$ ions which has recently becoming of significant technological importance for the fabrication of Cu interconnects on integrated circuit chips.

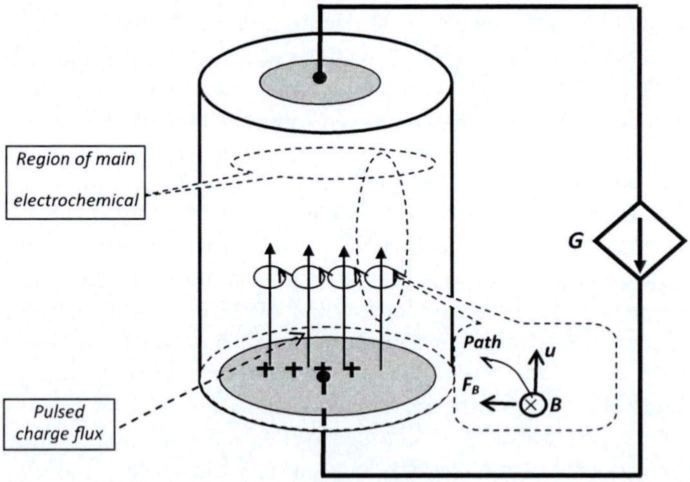

Fig. 4.7 Scheme of an electrochemical reactor: *G*—pulsed current generator, *u*—charge drift velocity, *B*—magnetic field density, F_B—Lorentz force

A simplified scheme of appliances being widely applied as electrochemical reactors and automated wafer processing equipment is presented in Fig. 4.7. Electrodes charged by supplying generator are able to carry out chemical reactions for realizations of electrocrystallization processes.

A variety of materials meeting the features in the art can be designed by choosing the type and composition of the electrolyte solution and the process parameters such as type and density of the current, temperature, electrolyte mixing, time duration of the process. In establishing the matrix of mechanisms that affect the electrochemical deposition process and its variation a great help can be obtained by applying the mathematical modeling approach. Development of a reaction model is an essential first step to obtaining a reactor model, which in turn is a suitable tool for process optimizations. It has to be noted that the electrochemical deposition process appears as a competition between nucleation and grain growth of the produced materials. In charging and discharging of a pulse ion flux, especially for short pulses, the double layer of the interface between electrolyte and cathode distorts the pulse current [10], and affects the over potential response acting on the electrolyte.

The results of the experimental electrochemical impedance spectroscopy [8] show that both the reactor current frequency and the magnetic flux density are important factors for efficiency of electrochemical reactors. The Lorentz force acting on the electrolyte in a double layer at electrode increases as the current density increases and improves the flow and conductivity of the electrolyte, what increases the efficiency of electrochemical deposition in terms of accuracy and smoothness of the surface of the layer. When the supplying current is time varying, the charged particles are attracted to each other under the influence of the Lorentz force, which causes shrinkage of the charged particle flux, i.e. the pinch effect appears. This effect leads to increase the density and pressure of ions. A current flowing through the electrolyte squeezes smaller cylinder, until the magnetic force does not balance the pressure. The magnetic properties of electrolytes and the interactions of internally induced magnetic fields on electrolyte properties and ionic transport characteristics have significant influence on the form of the mathematical model of the processes occurring in the electrochemical reactor. It should be based on the questions concerning the changes for mass, energy and momentum.

The instantaneous state of an electrochemical reactor usually depends on several variables and is described by one or several state functions of one or several variables. For instance, considering the electrode variables we must take into account the kind of the material, surface area, geometric and surface conditions. Such factors as mode transfer (diffusion, convection, migration, ...), surface concentrations and adsorptions have influences on the mass transfer variables. The temperature, pressure and time appear as external variables. Moreover, there are electrical variables such as the charge, current and potential. As solution variables we have the bulk concentrations of electroactive species, concentrations of other species and the kind of the solvent. It can be easily verified that the aspect of current distribution can have a major effect on the performance of electrochemical processes.

The derivation of a mathematical model taken from diffusion-migration transport equations (the Maxwell equations and mass balance equations) is an error prone and time-consuming tasks normally requiring specific expertise. Following appropriate procedure it is possible to select the variables that govern performance of the reactor. A start point leads to considering simulations of electrochemical depositions of nanostructures, and will then turn toward current–voltage relations to gain insights into the charge—electrolyte concentration and coupling mechanisms ultimately responsible for different pattern of the reactor output quantities and to identify the relationship between them. It should be emphasized that the circuit modelling not only aids manufacturers of nanostructural species in immediate returns on investment like increased yield and decreased consumable costs, but circuit models are essential to understand the mechanisms that enable of material and cause process variation, which affects device performance and process efficiency. Therefore, by imposing appropriate restrictions which are satisfied under corresponding electroplating conditions it is possible elaborate models which lead to satisfied results when matching them with experimental data. Such models have a two-fold purpose: first, they can account all the facts discovered experimentally, and second, they can be able predict the system behavior under various conditions of process realizations. Starting from this general imperative, the elaboration of a model for the interface phenomena can be performed by making a certain number of hypotheses which taken into account generally yield the governing equations. The representative circuit model predicts the capacitive and Faradic currents of the process, and with the same average and peak current density, the ramp up waveform has higher instantaneous peak current to charge transfer, which results in an improvement in the microstructure of the nanocomposites. The circuit model depicted in Fig. 4.8 makes it possible to quantify the evolution of the electrochemical reactor processes with respect to time.

Applying circuit laws we can establish the mathematical description of the above model and obtain

$$\begin{aligned}
\dot{x}_1 &= k_1\left(\frac{x_2 + x_3 - x_1}{R_e} + i_s\right),\\
\dot{x}_2 &= k_2\left(-\frac{x_2 + x_3 - x_1}{R_e} - (a + 3bx_4^2)x_2\right),\\
\dot{x}_3 &= k_3\left(-\frac{x_2 + x_3 - x_1}{R_e} - \frac{x_3}{R_p}\right),\\
\dot{x}_4 &= \dot{x}_2
\end{aligned} \tag{4.10}$$

Fig. 4.8 Circuit model of electrochemical deposition processes

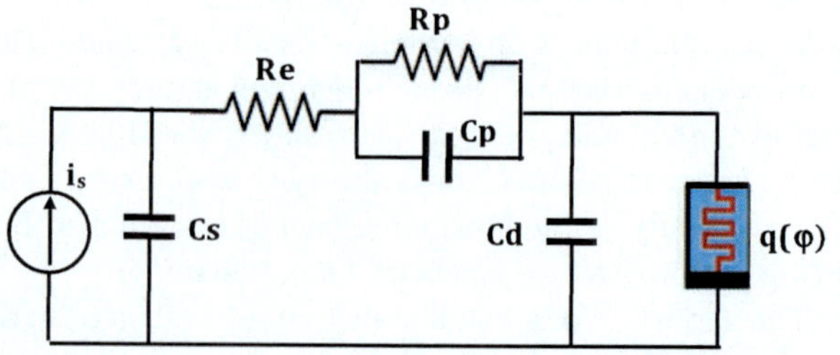

where $k_1 = 1/C_s$, $k_2 = 1/C_d$, $k_3 = 1/C_p$ and $x_4 = \varphi$. The dot is taken as a symbol for differentiation with respect to time. The charge–magnetic flux characteristic of the memristive element representing the relation between the current and its own magnetic field takes the form: $q(\varphi) = a\varphi + b\varphi^3$, where a and b denote constant parameters. The established circuit model can lead to methods for fast treating large quantities of electrocrystallizator data and extracting from them kinetic parameters importantly influenced the structure of the produced materials. Two types of periodic supplying currents play a significant role in the electrodeposition of nanostructured coatings: bipolar pulse current and sinusoidal rectified current. Both were taken into account in the sequel.

To examine the effect of various system parameters on the electrocrystallization process, we can perform numerical calculations applying a computer program MATLAB with using effective numerical integration procedures. Results of performed computer simulations of electrocrystallization processes with the focus on mixed mode oscillations (MMOs) as the dynamical switches between small amplitude oscillations (SAOs) and large amplitude oscillations (LAOs) are presented in the next section.

4.5 MMOs in Electrochemical Reactors

The modeling of the nucleation and material growth mechanism during the electrocrystallization process is extremely important due to its use in effective production of thin layers of pure metal or in the form of alloys and nanostructured composites with a high degree of precision. Generally, the manufacturing of nanostructures is too complicated for proper theory because the differential Eq. (4.10) are nonlinear, their analytical solution is often unknown and one has to use the computer to solve the problem numerically.

From the extensive set of solutions, in what follows the attention will be focused on the particular type of voltage and current waveforms that can be accordingly transformed on the effects of the electrochemical processes. The three basic modes of operation are illustrated in Fig. 4.9 which shows the solutions of (4.10) with $I \in \{0.15, 1.5, 15\}$ A, $a \in \{-1.25, -1.75, -2.75\}$, $b = 0.15$ and $T \in \{1,\ 2.5\}$s. In the SAOs only case, the small amplitude oscillations around the origin (0, 0, 0) are due to Hopf bifurcation for a = 0. In the LAOs only case, a trajectory passing close to the origin bypasses the region of small amplitude oscillations. The MMOs case is in some sense a combination of the previous two cases. The mechanism in which SAOs and LAOs occur is quite complex and has been the topics of [14–16]. In the MMOs case, a series of SAOs around the origin (considered canard solutions) undergoes a rapid canard explosion yielding an LAO, which, through a special return mechanism brings back the system into the vicinity of the origin. The canard explosion described in detail in [14] is triggered when a trajectory leaves a fold point of a cubic nonlinearity ending a series of SAOs and entering the relaxation mode with one or more LAOs. This explosion occurs, for example, in Fig. 4.9, bottom right part, when two or four SAOs

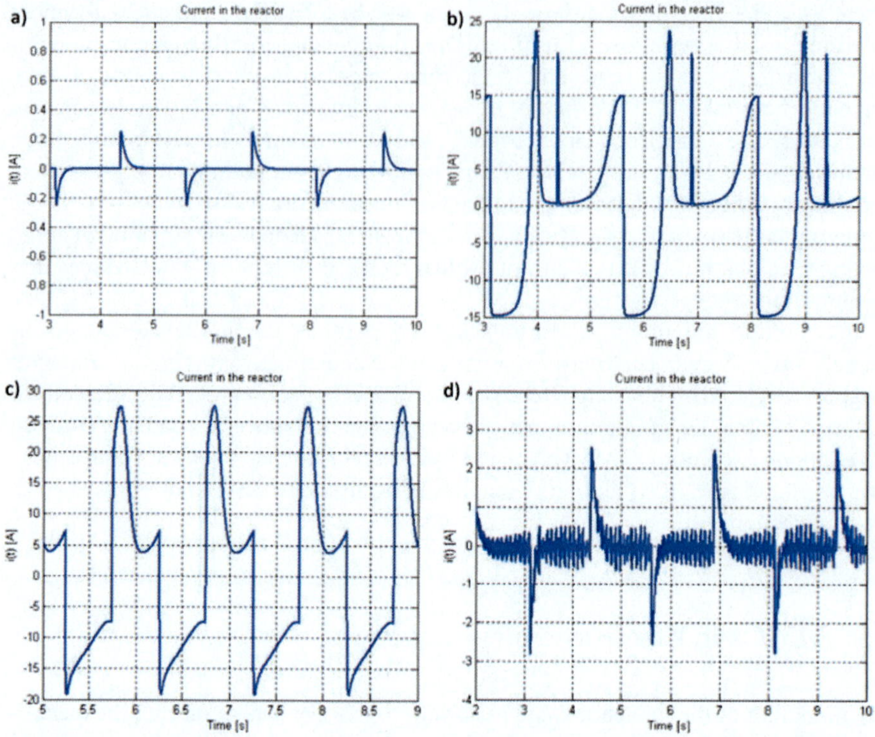

Fig. 4.9 Solutions of (4.10) for various sets of parameters I and b with unchanged other parameters: **a** SAOs at $I = 0.15$ A, $a = -1.75$ and $b = 0.15$, **b** LAOs at $I = 15$ A, $a = -2.75$ and $b = 0.15$, **c** LAOs at $I = 15$ A, $a = -1.25$ and b $= 0.15$, **d** MMOs at $I = 1.5$ A, $a = -1.75$, $b = 0.15$ and $T = 2.5$ s

transform into an LAO. Depending on the parameters, the system may continue with one (or more) LAOs, or may go through a new series of SAOs after which trajectory leaves again the vicinity of the origin (a fold) and the phenomenon repeats.

Taking into account a small amount of perturbation in the characteristic $q(\varphi)$ it is possible to get MMOs exhibiting in each period a train of SAOs separated by a large amplitude pulse (Fig. 4.10a). Varying pairs of control parameters, one can observe how the dynamical behaviour of the system changes and map out "dynamical phase diagrams" showing regions of the parameter space in which qualitatively different MMOs occurs, or, by varying a single parameter, one can trace out hysteresis loops of the sort shown in Fig. 4.10b. This type of time-varying electric currents and magnetic fields are most probably generated by an electric double layer within individual particles and a temporal excess of ion carriers within the sample. Perhaps the shape and magnitude of the induced temporal signals depend on the reaction zone propagation mode, reaction mechanism, and reactant properties. Therefore, by imposing appropriate restrictions, which are satisfied under corresponding electroplating conditions, it is possible elaborate models which lead to satisfied results when matching them

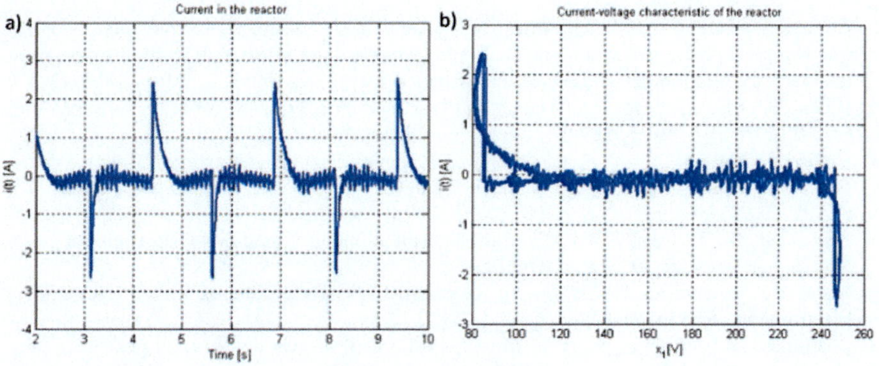

Fig. 4.10 MMOs at $I = 1.5$A (bipolar pulse), $b = 0.5$ and $T = 2.5$: **a** reactor current, **b** voltage–current characteristic of the reactor

with experimental data. This is of great importance in the case of manufacturing nanostructured materials whose final structure is very sensitive to rapid changes in the conditions of the manufacturing process. A particularly interesting context is that of current waveforms supplying the electrocrystallizator because their influences on mixed mode oscillations seem most important for the efficiency of realized processes.

References

1. Silberberg, M.: Chemistry—The Molecular Nature of Matter and Change. McGraw-Hill Science/Engineering/Math, New York (2008)
2. Epstein, I.R., Pojman, J.A.: An Introduction to Nonlinear Chemical Dynamics: Oscillations, Waves, Patterns, and Chaos. Oxford University Press, New York (1998)
3. Laganà, A., Gregory, A.P.: Chemical Reactions. Basic Theory and Computing. Springer, New Yok (2018)
4. Trzaska, M., Trzaska, Z.: Nanomaterials produced by electrocrystallization method. In: Aliofkhazraei, M., Makhlouf, A. (eds.) Handbook of Nanoelectrochemistry. Springer, Cham (2016)
5. Lona, F., Maria, L.: Step by Step Approach to the Modeling of Chemical Engineering Processes. Using Excel for Simulation. Springer, New Yok (2018)
6. Kreysa, G., Ota, K.-I., Savinell, R.F. (Eds.): Encyclopedia of Applied Electrochemistry. Springer, New York (2014)
7. Cieślak, G., Trzaska, M.: Tribological properties of nanocomposite Ni/graphene coatings produced by electrochemical reduction method. Compos.: Theory Pract. **2**, 79–83 (2016)
8. Trzaska, M., Trzaska, Z.: Electrochemical impedance spectroscopy in materials engineering. Publisher Office of the Warsaw University of Technology, Warsaw (2010)
9. Sangeetha, S., Kalaignan, G.P., Anthuvan, J.T.: Pulse electrodeposition of self-lubricating Ni–W/PTFE nanocomposite coatings on mild steel surface. Appl. Surf. Sci. **359**, 412–419 (2015)
10. Kamnerdkhag, P., Free, M.L., Shah, A., Rodchanarowan, A.: The effects of duty cycles on pulsed current electrodeposition of ZneNieAl$_2$O$_3$ composite on steel substrate: microstructures, hardness and corrosion resistance. Int. J. Hydrogen Energy **42**, 20783–20790 (2017)

11. Abdul, J., Yasin, G., Khan, W.Q., Anwar, M.Y., Korai, R.M., Nizam, M.N., Muhyodin, G.: Electrochemical deposition of nickel graphene composite coatings: effect of deposition temperature on its surface morphology and corrosion resistance. Royal Soc. Chem. **7**, 31100–31109 (2017)
12. Djokić, S.S.: Electrodeposition: Theory and Practice. Springer, New York (2010). https://doi.org/10.1007/978-1-4419-5589-0
13. Sagués, F., Epstein, I.R.: Nonlinear chemical dynamics. Dalton Transactions (March 2003)
14. Focus issue: Mixed-mode oscillations: experiment, computation, and analysis. Chaos, 18, 015101 (2008)
15. Marszalek, W., Trzaska, Z.W.: Mixed-mode oscillations in a modified Chua's circuit. Circ. Syst. Signal Proc. **29**, 1075–1087 (2010)
16. Marszalek, W., Trzaska, Z.W.: Properties of memristive circuits with mixed-mode oscillations. Electron. Lett. **51**(2), 140–141 (2015)
17. Beardmore, R.E., Laister, R.: The flow of a differential-algebraic equation near a singular equilibrium. SIAM J. Matrix Anal. **24**(1), 106–120 (2002)
18. Goldbeter, A.: Biochemical Oscillations and Cellular Rhythms: The Molecular Bases of Periodic and Chaotic Behaviour. Cambridge University Press, Cambridge, UK (1996)
19. Inaba, N., Kousaka, T.: Nested mixed-mode oscillations. Physica D **401**, 132–152 (2020)
20. Estevez-Rams, E., Estevez-Moya, D., Aragon-Fernandez, B.: Phenomenology of coupled nonlinear oscillators. CHAOS **28**, 023110 (2018)
21. Willy, Ch., Neugebauer, E.A.M., Gerngroß, H.: The concept of nonlinearity in complex systems. Eur. J. Trauma **200**(1),·11–22
22. Rachwalska, M.: Mixed mode, sequential and relaxation oscillations in the Belousov-Zhabotinsky system. Zeitschrift Naturforschung **62a**, 41–55 (2007)
23. Lynch, D.T., Rogers, T.D., Wanke, S.E.: Chaos in a continuous stirred tank reactor. Math. Model. **3**, 103–116 (1982)
24. Lapicque, F.: Electrochemical reactors. In: Chemical Engineeering and Chemical Process Technology, vol. III. Encyclopedia of Life Support Systems (2011)
25. Feng, J.M., Gao, Q.Y., Li, J., Liu, L., Mao, S.C.: Current oscillations during the electrochemical oxidation of sulfide in the presence of an external resistor. Sci. China Ser. B: Chem. **51**(4), 333–340 (2008)
26. Green, B.J., Wang, W., Hudson, J.L.: Chaos and spatiotemporal pattern formation in electrochemical reactions. Forma **15**, 257–265 (2000)
27. L. M.A. Monzon, J.M.D. Coey, Magnetic fields in electrochemistry: The Lorentz force. A mini-review, Electrochemistry Communications, 2014, 1–13
28. Sulaymon, A.H., Abbar, A.H.: Scale-up of electrochemical reactor. In: Electrolysis, Chapter 9. InTech, Rijeka (2011)
29. Białostocka, A.M.: The electrochemical copper structure forming in the presence of the magnetic field. Electrotech. Rev. **89**(10), 254–256 (2013)
30. Drews, T.O., Krishnan, S., Alameda, J.C., Jr., Gannon, D., Braatz, R.D., Alkire, R.C.: Multiscale simulations of copper electrodeposition onto a resistive substrate. IBM J. Res. Dev. **49**(1), (2005)

Chapter 5
Mixed-Mode Oscillation Synchronization in Coupled Oscillators

5.1 Introduction

Numerous important physical and engineering systems exhibit coupled oscillators connected in such a way that energy can be transferred between them. Then there are interactions amongst the different oscillating components of a given system. For instance, atoms oscillate around their equilibrium positions, and the interaction between atoms is responsible for the coupling of their instantaneous displacements. Examples from biology show synchronous flashes of fireflies, the chirping of crickets, pacemaker cells, or the human nerve cluster responsible for circadian rhythm [1–5]. The dynamics of coupled oscillators are typically complex and need not be periodic. All mechanical and electrical oscillation problems restrict, in the case of small amplitude vibrations, to problems involving one or several pairs of coupled oscillators. Problems referring to vibrations of strings, membranes, and elastic solids, electrical or acoustic devices can be reduced to problems of coupled oscillators. Moreover, many physical, chemical, biological and technical systems should be modeled using interacting nonlinear oscillators [4].

An inspiring quality of such systems is their ability to oscillate jointly at a common frequency or phase even when the individual oscillators exhibit different self frequencies or initial phases. In dynamical systems investigations this phenomenon is known as synchronization. Usually, small groups of oscillators are directly subject to synchronizing systems, such as lasers that are coupled to increase power output [2], and nanomechanical oscillators, where synchronization to a common frequency can eliminate the inevitable differences in the response frequency appearing from imperfections in their manufacturing. In particular, a problem of synchronization of same structure oscillators is of great importance because such systems are widespread in studies of neuronal dynamics, nonlinear optics, biophysics, geophysics, telecommunication and information engineering, economics, and ecology [5]. Moreover, the phenomena generated by slow–fast and multiple-time-scale dynamics, such as canards and MMOs, have attracted the attention of many researchers and practitioners

Z. Trzaska, *Mixed Mode Oscillations (MMOs)*, Studies in Systems, Decision and Control 374, https://doi.org/10.1007/978-3-030-76867-6_5

recently [6]. MMO synchronization generated by coupled slow—fast and multiple-time-scale oscillators is a very important subject that extends canard synchronization. It can be expected to become a crucial problem in investigations of slow-fast and multi-scale dynamics in the near future. Studies of systems synchonization from physics and chemical, biological and hydraulic processes encompass, between others, the Furuta pendulum [7], laser arrays [8], Josephson junctions [9], enzymatic reactions, and fluid systems [10].

Another possible field of application of the oscillator synchronizations is the support of medical diagnosis, e.g., cell counting in large data sets of microscopic histological images [5] or object detection in cross-section of 3D magnetic resonance images (MRI) data [11].

In the sequel the synchronization of mixed-mode oscillations (MMOs) generated by two identical weakly coupled electric oscillators, each of which can exhibit MMOs and a succession of MMO-incrementing bifurcations are investigated. Coupled nonlinear oscillators give rise to important dynamics and can been used to investigate synchronous phenomena such as oscillations in neural networks and particularly in FitzHugh–Nagumo dynamics.

5.2 Complete Synchronization of Mixed-Mode Oscillations in Weakly Coupled Electric Oscillators

In a wide range of MMO-generating dynamics, including chemical reactions [1–3, 12] and electrical circuits [13, 14] with three or more state variables, counting "quadratic" or "cubic" feedback mechanisms, complex oscillatory behavior and chaos can emerge. Realized up-to-date studies showed that complex phenomena such as birhythmicity and chaos could arise by coupling two instability-generating processes. For instance, disruption of normal heart rhythms increases cardiovascular disease risk. The complex oscillatory behavior observed in many nonlinear dynamical systems can be explained in terms of the fast and slow variables which have of considerable meaning in studying specific chemical reactions and biological systems. Ever since it was found that significant MMOs can occur in extended slow–fast ordinary differential equations (ODEs) that can generate canards [4–6, 11], they are still the subject of intense research [4, 5, 7–10, 13, 15–26].

Analyzing numerically extended slow–fast or multiple-time-scale solutions of ODEs that can generate canards, it is frequently to encounter MMOs, which could be universal phenomena such as period-doubling (PD) bifurcations or Arnold tongues. For these, among others, reasons MMOs are now seen as generalized canard phenomena [4], so that MMO synchronization generated by coupled slow–fast and multiple-time-scale oscillators is a very important problem that extends canard synchronization. Synchronization of both extrinsic and intrinsic rhythms is crucial for optimal functioning of the heart, and represents one important method in which the full chronotropic range of the heart is achieved. Desynchronization of rhythms

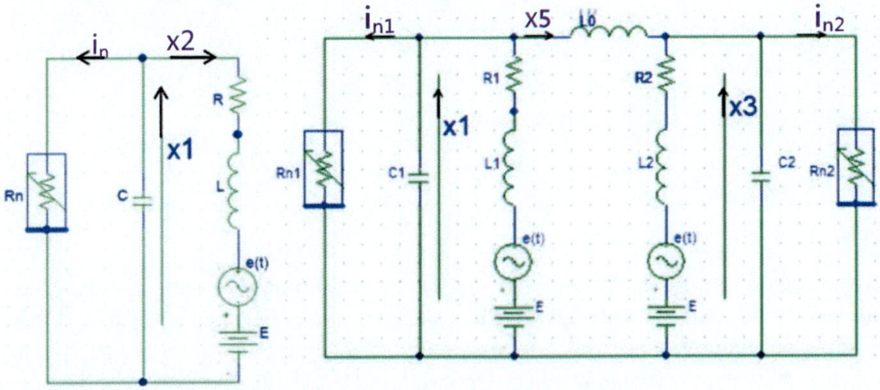

Fig. 5.1 Schemes of circuits susceptible to the generation of MMOs in: **a** a separate circuit, **b** circuits coupled with a high inductance coil

leads to pathological phenotypes, including cardiac arrhythmias, and also because a similar set of molecules are involved in atrial and ventricular myocytes as in SAN cells. In such a case the desynchronization can lead to the heart failure phenotype. Hence, one should expect that this will become an important problem in investigations of slow–fast and multiple-time-scale dynamics in the future [1, 27].

Here, we consider dynamical circuits generating MMOs taking into account two nonlinear oscillators coupled by an inductor with large inductance. This means that the two oscillators are weakly coupled [28].

Figure 5.1a shows a diagram of the concerned circuit as a basic structure for a driven nonlinear dynamical oscillator. If the capacitance C is small then the voltage x_1 changes rapidly in time. The characteristics of the nonlinear resistor R_n is determined by

$$i_n = ax_1 + bx_1^3 \tag{5.1}$$

with constants $a < 0$ and $b > 0$. Application of the Kirchhoff current and voltage laws yields

$$\begin{aligned} \dot{x}_1 &= \propto (-x_2 - ax_1 - bx_1^3) \\ \dot{x}_2 &= \beta(x_1 - Rx_2 - e - E) \end{aligned} \tag{5.2}$$

where $\alpha = C^{-1}$ and $\beta = L^{-1}$. The remaining markings are the same as in Fig. 5.1a. The dot corresponds to differentiation with respect the time.

By introducing substitutions

$$\tau = \frac{t}{aL}, c = \frac{aL}{C}, k = aR, \ \omega = aL\omega_1,$$

$$A_0 = \sqrt{\frac{b}{a}}\,E,\; A_1 = \sqrt{\frac{b}{a}}E_1,\; x = \sqrt{\frac{b}{a}}\,\mathrm{x}_1,\; y = \sqrt{\frac{b}{a}}\,x_2 \tag{5.3}$$

the above relations take a dimensionless form

$$\begin{aligned} x' &= c\left(-y - x - x^3\right) \\ y' &= x - ky - A_0 - A_1 \sin(\omega\tau) \end{aligned} \tag{5.4}$$

Here, the sign represents Newton's notation for differentiation, that is, d/dτ. The state variables x and y correlate with the voltage x_1 across the capacitor C and the current x_2 through the inductor L, respectively. The parameter values are such that both a stable focus and stable relaxation oscillations arising from an unperturbed subcritical Hopf bifurcation can be investigated. Computer simulations have been performed with applying program MATLAB for the set of element parameters $A_0 = 0.207$, $A_1 = 0.0105$, $k = 0.9$, c = 10.00, $\omega = 0.58$, and initial conditions $x(0) = 1.12847365485316$, $y(0) = 0.30858426904279$. The initial conditions are chosen on the path x´ = 0, and x″ < 0. Since there is only one attractor, this is why MMO-incrementing bifurcations have not emerged. Results of computer simulations are shown in Fig. 5.2.

The dynamical circuit shown in Fig. 5.1b forms a synchronized oscillator, in which two identical driven nonlinear oscillators are coupled by an ideal inductor with large inductance, meaning that they are weakly coupled. Following Kirchhoff's laws and applying the previous procedure for parameters and variables substitutions, the circuit's governing equation can be represented by the succeeding system of five dimensionless ODEs

$$\begin{aligned} \dot{x}_1 &= c(-x_2 - x_1 - x_1^3 - x_5 \\ \dot{x}_2 &= x_1 - k_2 x_2 - A_0 - A_1 \sin(\omega\tau) \\ \dot{x}_3 &= c\left(-x_4 - x_3 - x_3^3 + x_5\right) \\ \dot{x}_4 &= x_3 - k_4 x_4 - A_0 - A_1 \sin(\omega\tau) \\ \dot{x}_5 &= d(x_2 - x_3) \end{aligned} \tag{5.5}$$

where variables and parameters with indices 1 and 2 correspond to left hand side oscillator while these with indices 3 and 4 correspond to right hand side oscillator. The variable x_5 refers to the current in the coupling coil. The L/L_0 ratio is denoted by d. Throughout this study the intensity of the coupling between the two identical oscillators was set to 0.01.

Two distinctly different types of MMOs can be observed in Fig. 5.3, namely, completely and incompletely synchronized MMOs, what is illustrated by Fig. 5.3a, b, respectively. These results indicate that MMOs, which are periodic, are always synchronized, either completely or incompletely. However when the coupling is strong the oscillators are still synchronized although the MMOs no longer appear. They are then replaced by relaxation oscillations (Fig. 5.3b). Moreover, complete

MMOs synchronization is very sensitive to the initial conditions. For instance only little change in $x_2(0)$ results in important variations of the oscillator response (Fig. 5.4).

Results of performed simulations indicate that a succession of MMO-incrementing bifurcations (MMOIBs) also known as complex period-adding sequences can generate completely synchronized MMOs (Fig. 5.5a). The mechanism responsible for these successive MMOIBs can be explained using a modified Lorenz plot [29–31]. Although MMOs are periodic, the bifurcation structures can be precisely observed because the MMOs generated by MMOIBs can have high periods. Let the time τ at which the solution enters Region b from Region 1 be denoted by τ_n. Figures 5.5b show the plots, which represent the relation of $\omega\tau_n$ and $\omega\tau_{n+1}$ for large *n*. Denoting by τ_n the time τ at which the solution enters region *B* from region *A* gives the relation of $\omega\tau_n$ and $\omega\tau_{n+1}$ for large *n*.

The plot of Fig. 5.5b shows that MMOIBs are generated in a similar mechanism that causes period-adding phenomena in the circle map, which strongly suggests that MMOIBs occur successively many times. It is worth noticing that the coupled nonlinear oscillators exhibit MMOs and successive MMOIBs near subcritical Hopf bifurcations when driven by weak periodic forcing sources.

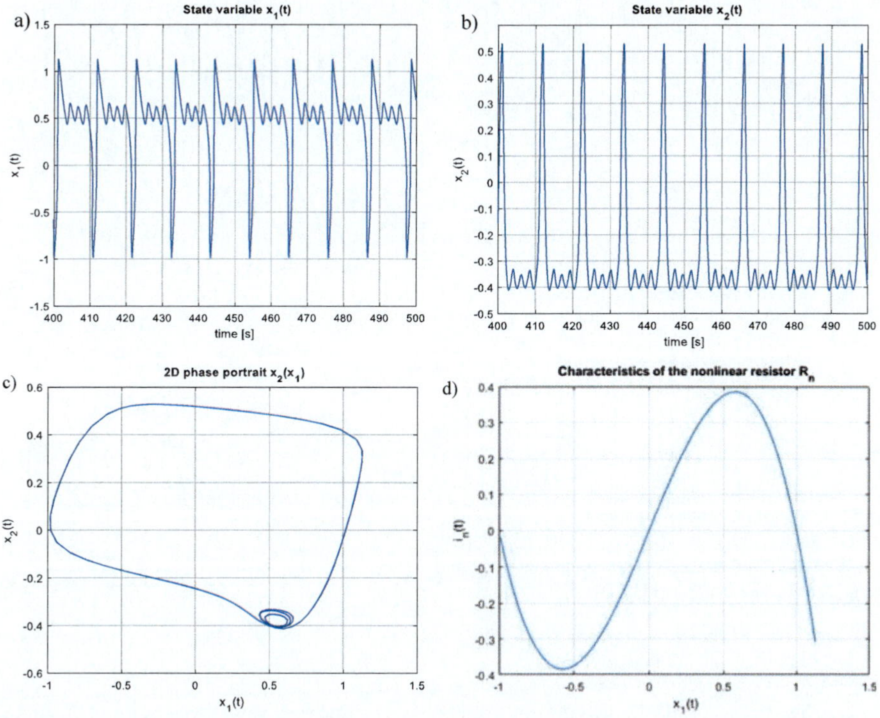

Fig. 5.2 Simulation results of a separate oscillator: **a** state variable $x_1(t)$, **b** state variable $x_2(t)$, **c** limit cycle, **d** characteristics of nonlinear resistor R_n

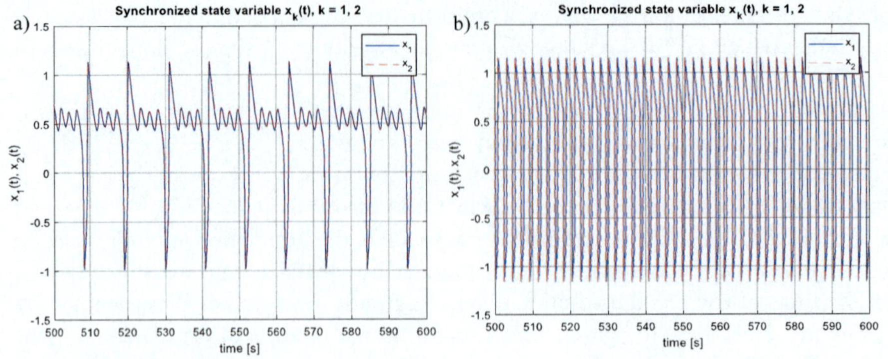

Fig. 5.3 Time series responses of the synchronized oscillators: **a** weak coupling (b = 10), **b** strong coupling (b = 1000)

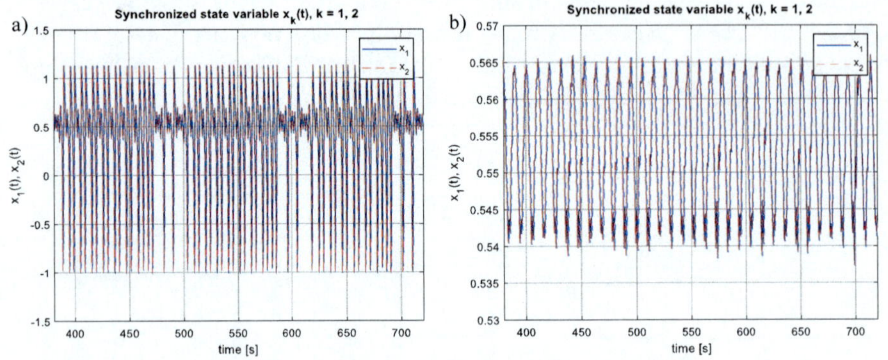

Fig. 5.4 Illustration of sensitivity to the initial conditions: **a** $x_5(0) = 10^{-4}$, **b** $x_5(0) = 10^{-3}$

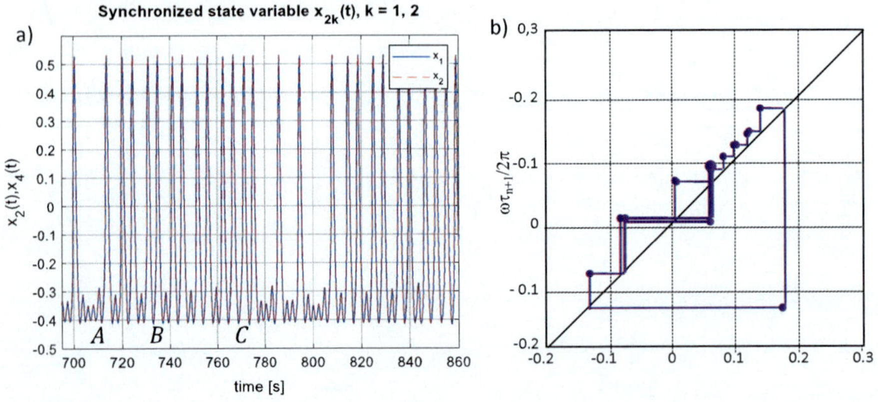

Fig. 5.5 Succession of MMO-incrementing: **a** time series of the state variable $x_2(t)$, **b** Lorenz plot

5.3 Synchronization of Oscillations in the Conduction System of the Human Heart

Human heart rate is not constant in long time period. In real, heart rate variability is a substantial factor for the effective functioning of the cardiovascular system and a significant factor in clinical diagnosis. Recently, the systematic investigation into have been focused on cardiovascular system responses to changes caused by various diseases affecting malfunction in the heart to a point where it can no longer pump enough blood to the human body. All such studies of the available materials and sources were carried out in order to precisely establish the facts and draw new conclusions. Many patients with dilated cardiomyopathy and impaired systolic function are burdened by dyssynchronous activation of portions of the left or right ventricle such that the mechanical advantage of simultaneous contraction of the ventricular walls is lost. This loss of intraventricular synchrony causes the heart to work inefficiently, resulting in decreased stroke volume and reduced cardiac output while simultaneously increasing the work of the heart. In contrast with pharmacologic therapies, the cardiac resynchronization therapy strives to alter the dyssynchrony within the left ventricle, between the left and right ventricles, and between the atria and ventricles. The increasing adoption of biventricular pacing as a means to reach interventricular, intraventricular, and atrio-ventricular synchronization provides clinicians a powerful nonpharmacologic instrument in the treatment of congestive heart failure. It should be emphasized that the instantaneous state of any element of each cardiovascular system depends on multiple variables and can be described by one or several state functions of appropriately selected quantities. One of the most important organs in human body is heart which pumps blood throughout the body using blood vessels [32–35].

The source of such action is not fully known on an ongoing basis and remains an open research topic. However, it is well known that the autonomic nervous system eases the heart rate of every human being. One way to interpret how this happens is to build right models.

A model with coupled nonlinear oscillators of the conduction system of the heart plays an essential role to determine the correct diagnosis and administer high-quality treatment cardiac failures. The conducting system of the human heart: the sinoatrial (SA) node (the primary pacemaker), the atrioventicular (AV) node and the His-Purkinje system—could be treated as a network of self excitatory elements (Fig. 5.6). These elements may be modeled as interacting nonlinear oscillators [36]. The AV node is considered as an active pace-maker and allows to designate a model of two nonlinear coupled oscillators in order to describe the interaction between the SA and the AV nodes. These two nonlinear oscillators are based on a modification of the van der Pol oscillator, so that the generated waveforms resemble the action potentials of cells in the SA and the AV node, respectively [37–39].

The phenomenological approach using nonlinear oscillators allows a global analysis of heartbeat dynamics by investigating interactions between the elements of the system. A unidirectional coupling between the SA and AV nodes plays a primary

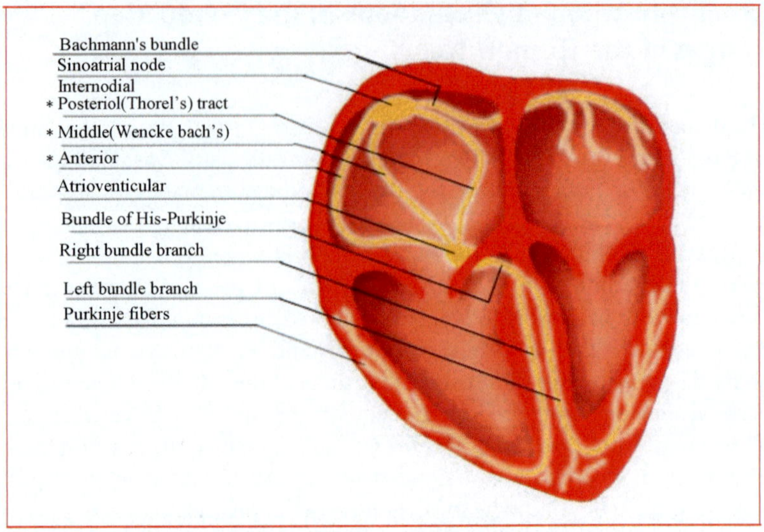

Fig. 5.6 Longitudinal section of the structure of the human heart

role. Its significance lies in the fact that it is able to reproduce certain recognizable phenomena that occur in a clinically recorded human heart rate: irregular heart rate, asystole, sinus pause, the vicious paradox, some types of heart block, and others. To study the properties of cardiac tissue both FitzHugh-Nagumo and Hodgkin-Huxley type models are used. The most important of these properties are: the shape, properties of the refractory period, and modes of pulse frequency shifting. They can be reproduced using ion channel models [12, 40].

Under physiologic conditions, the AV junction is traditionally regarded as a passive conduit for the conduction of impulses from the atria to the ventricles. An alternative view, namely that subsidiary pacemakers play an active role in normal electrophysiologic dynamics during sinus rhythm, has been suggested based on nonlinear models of cardiac oscillators. The structures introduced into the phase space of the modified van der Pol system by some kind of Duffing terms yield well the conduction processes between SA and AV nodes in the heart. Modifying the relaxation term leads to an oscillator [1] that reproduces two of the three ways that a node of the conduction system of the heart may change its frequency (Fig. 5.7).

Two coupled oscillators in the area of node SA or alternatively node AV may be described by equations

$$
\begin{aligned}
\frac{dx_1}{dt} &= x_2 \\
\frac{dx_2}{dt} &= \propto_1 (x_1 - a_{12})(x_1 - a_{12})x_2 - c_1(x_1 - d_{11})(x_1 - d_{12})x_1 - e \\
\frac{dx_3}{dt} &= x_4 \\
\frac{dx_4}{dt} &= \propto_2 (x_{31} - a_{21})(x_3 - a_{22})x_4 - c_2(x_3 - d_{21})(x_3 - d_{22})x_3 + e
\end{aligned}
\tag{5.6}
$$

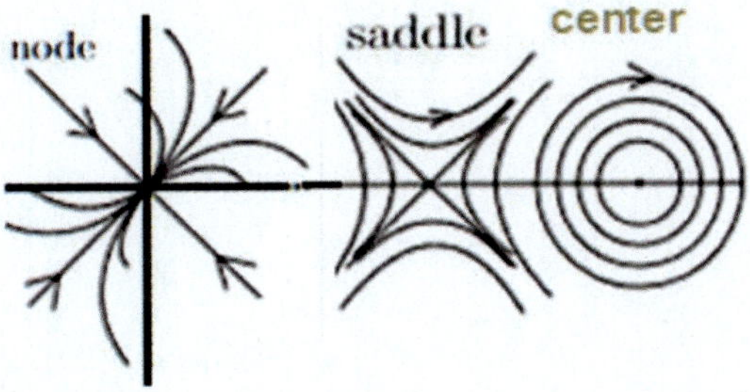

Fig. 5.7 Scheme of the trajectories of nonlinear oscillators in the phase space

where x_1 and x_3 denote the potentials. The currents acting on the nodes are marked by x_2 and x_4, respectively. For $k = 1, 2$ parameters α_k reflect the damping. Variable $e = e(t)$ denotes the signal of synchronization. All parameters and variables in this section are dimensionless but the period of the oscillations was maintained in a range similar to that observed in clinical practice. The established model exhibits three fixed points: a single unstable center (at $\boldsymbol{x} = \mathbf{0}$), a stable node, and a hyperbolic saddle. To preserve the adequate ratio of the frequencies of the two nodes, to model the SA node was used $d_{12} = 12$ and to model the AV node the setting $d_{21} = 7$ has been applied. Note that in the phase space these parameters define the position of the stable nodes. The position of the saddles are fixed by d_{11} and d_{21}, respectively. Meanings of all other parameters in Eq. (5.6) are clear from the structures of particular components. The model reproduces several phenomena well known in cardiology, such as certain properties of the sinus rhythm and heart block. In particular, the model reproduces the decrease of heart rate variability with an increase in sympathetic activity. Two such oscillators coupled unidirectionally and asymmetrically, allow us to reproduce the properties of heart rate variability obtained from patients with different kinds of heart block including sinoatrial blocks of different degree and a complete AV block [36, 41–46].

The conduction system of the heart is being managed by the parasympathetic nervous system through such neurotransmitters as acetylcholine having a very fast rise time and quickly decays. This effect can be verified by applying a short rectangular pulse of potential in the model (5.6). In computer simulation the following parameters $\alpha = -5$, $\beta = 3$, $a_1 = 1$, $a_2 = -1$, and $d_{11} = d_{21} = 2.5$, $d_{12} = 7$, $d_{22} = 12$ were taken into account. Applying the external pulse at the end of the restive phase (Fig. 5.8a, c) leads to the increased resultant instantaneous period of oscillations. The period of the oscillation is longer than normal one because the excitation decreases momentarily the resting potential. If the shape of the applied external pulse is chosen respectively a long pause sustaining several normal periods of the oscillation may occur. This corresponds to similar case which may appear in the human heart as a result of a single supraventricular ectopic beat [25]. It should be underlined that positive external single pulses generally do not have such a potent effect on the instantaneous period of the oscillator. It comes from the symmetry breaking effect of the hyperbolic saddle located on one side of the limit cycle only.

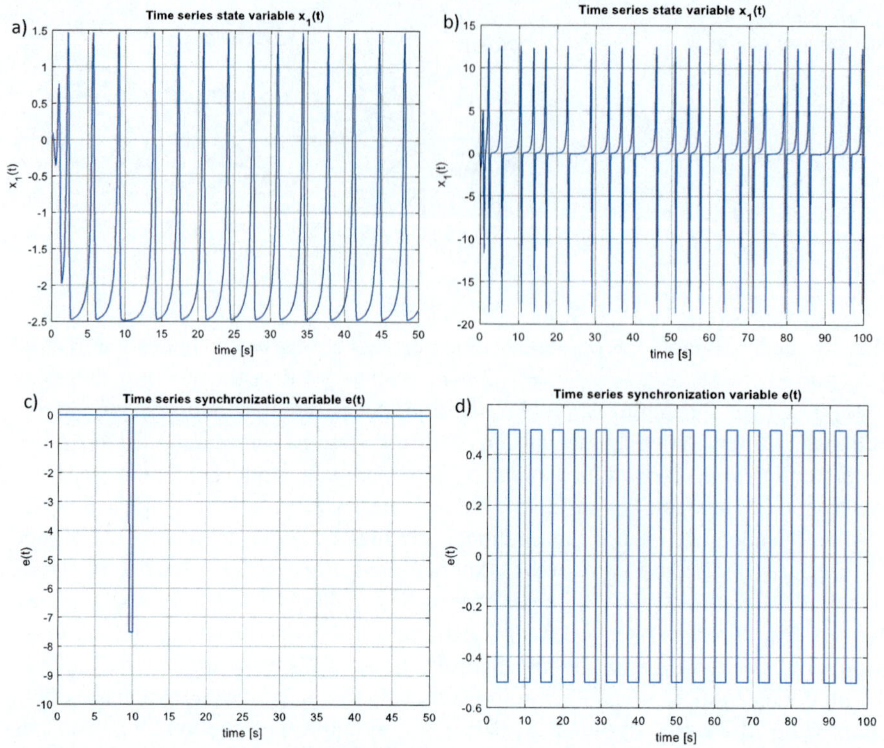

Fig. 5.8 Simulation results of a synchronization effect due to the application of pulses of the appropriate value and phase: **a** potential $x_1(t)$ synchronized with a single negative pulse $e(t)$ **b** potential $x_1(t)$ synchronized with a string of periodic bipolar pulses $e(t) = e(t + T)$, **c** single negative pulse $e(t)$, **d** string of periodic bipolar pulses $e(t) = e(t + T)$

Figure 5.8b presents the effect of the application of a string of periodic bipolar pulses with a relatively modest frequency (Fig. 5.8d). It can be easily identified that some of the pulses have been replaced by subthreshold oscillations. In result the effective period of the oscillations has been substantially increased. The number of subthreshold oscillations between the normal pulses of the oscillator may be controlled by setting the amplitude and period of the bipolar pulse perturbation. This property of the established model admits reconstructing not only the effect of acetylocholine bursts but also that of a fast parasystole. Remembering is essential that the amplitude of the perturbing bipolar pulses may not exceed the critical value at which the trajectory of the oscillator will be flipped across the saddle onto the stable node and the oscillations will be suppressed completely. A string of bidirectional periodic pulses with a frequency smaller than that of the oscillator changes the individual phases of the oscillator responses.Therefore the presented simple nonlinear model can help to improve the understanding of the complex phenomena involved in heart rhythm generation as well as of heart rate control and function.

References

1. Żebrowski, J.J., Grudziński, K., Buchner, T., Kuklik, P., Gac, J., Gielerak, G., Sanders, P., Baranowski, R.: A nonlinear oscillator model reproducing various phenomena in the dynamic of the conduction system of the heart, to appear in Chaos 17, Focus Issue "Cardiovascular Physics" (2007)
2. Grudziński, K., Żebrowski, J.J.: Modeling cardiac pacemakers with relaxation oscillators. Physica A **336**, 153–152, (2004)
3. Bub, G., Glass, L.: Bifurcations in a discontinuous circle map: a theory for a chaotic cardiac arrhythmia. Int. J. Bifurcations Chaos **5**, 359–371 (1995)
4. di Bernardo, D., Signorini, M.G., Cerutti, S.: A model of two nonlinear coupled oscillators for the study of heartbeat dynamics. Int. J. Bifurcations Chaos **8**, 1975–1985 (1998)
5. Thompson, J.M.T., Steward, H.B.: Nonlinear Dynamics and Chaos. Wiley, New York (2002)
6. Keener, J., Sneyd, J.: Mathematical Physiology, Interdisciplinary Applied Mathematics 8. Springer, New York (1998)
7. Katholi, C.R., Urthaler, F., Macy, J., Jr., James, T.N.: A mathematical model of automaticity in the sinus node and the AV junction based on weakly coupled relaxation oscillators. Comp. Biomed. Res. **10**, 529–543 (1977)
8. Honerkamp, J.: The heart as a system of coupled nonlinear oscillators. J. Math. Biol. **18**, 69–88 (1983)
9. Zhang, H., Holden, A.V., Kodama, I., Honjo, H., Lei, M., Varghese, T., Boyett, M.R.: Mathematical models of action potentials in the periphery and center of the rabbit sinoatrial node. Am. J. Physiol. Heart Circ. Physiol. **279**, H397–H421 (2000)
10. West, B.J., Goldberger, A.L., Rovner, G., Bhargava, V.: Nonlinear dynamics of the heartbeat. The AV junction: passive conduict or active oscillator? Phys. D **17**, 198–206 (1985)
11. van der Pol, B., van der Mark, J.: The heartbeat considered as a relaxation oscillation and an electrical model of the heart. Phil. Mag. **6**, 763–775 (1928)
12. Grudziński, K., Żebrowski, J.J., Baranowski, R.: A model of the sino-atrial and the atrio-ventricular nodes of the conduction system of the human heart. Biomed. Eng. **51**, 210–214 (2006)
13. Hoffman, B.F., Cranefield, P.F.: Electrophysiology of the Heart. Mc Graw Hill, New York (1960)
14. Lewis, T.J., Keener, J.P.: Wave-block in excitable media due to regions of depressed excitability. SIAM J. Appl. Math. **61**, 293–316 (2000)
15. FitzHugh, R.: Impulses and physiological states in theoretical models of nerve membrane. Biophys. J. **1**, 445–466 (1961)
16. Postnov, D., Kee, H., Seung, K., Hyungtae.: Synchronization of diffusively coupled oscillators near the homoclinic bifurcation. Phys. Rev. E **60**, 2799–2807 (1999)
17. Jalife, J., Slenter, V.A., Salata, J.J., Michaels, D.C.: Circ. Res. **52**, 642–656 (1983)
18. Demir, S.S., Clark, J.W., Giles, W.R.: Am. J. Physiol. Heart. Circ. Physiol. **276**, H2221–H2244 (1999)
19. Bub, G., Glass, L.: Bifurcations in a discontinuous circle map: a theory for a chaotic cardiac arrhythmia. Int. J. Bifurcation Chaos Appl. Sci. Eng. **5**, 359–371 (1995)
20. di Bernardo, D., Signorini, M.G., Cerutti, S.: A model of two nonlinear coupled oscillators for the study of heartbeat dynamics. Int. J. Bifurcation Chaos Appl. Sci. Eng. **8**, 1975–1985 (1998)
21. van der Pol, B., van der Mark, J.: The heartbeat considered as a relaxation oscillation and an electrical model of the heart. Philos. Mag. **6**, 763–775 (1928)
22. West, B.J., Goldberger, A.L., Rovner, G., Bhargava, V.: Nonlinear dynamics of the heartbeat. The AV junction: passive conduit or active oscillator? Physica D **17**, 198–206 (1985)
23. Katholi, C.R., Urthaler, F., Macy, J., Jr., James, T.N.: Mathematical model of automaticity in the sinus node and the AV junction based on weakly coupled relaxation oscillators. Comput. Biomed. Res. **10**, 529–543 (1977)
24. Grudziński, K., Żebrowski, J.J.: Modeling cardiac pacemakers with relaxation oscillators. Phys. A **336**, 153–162 (2004)

25. Theory of Heart: edited by L. P. Hunter, and A. McCulloch, Springer, Berlin, Glass (1990)
26. In: Zipes, P.D., Jalife, J. (eds) Cardiac Electrophysiology: From Cell to Bedside. Saunders, Philadelphia (2000)
27. Buchner, T., Gielerak, G., Żebrowski, J.J.: Proceedings of the European Study Group for Cardiovascular Oscillations ESGCO, Jena 2006, pp. 191–194 (2006)
28. Joyner, R.W., van Capelle, F.J.L.: Propagation through electrically coupled cells: how a small SA node drives a large atrium. Biophys. J. **50**, 1157–1164 (1986)
29. Glass, L.: Synchronization and rhythmic processes in physiology. Nature, London **410**, 277–284 (2001)
30. Kotani, K., Takamasu, K., Ashkenazy, Y., Stanley, H.E., Yamamoto, Y.: A model for cardio-respiratory synchronization in humans. Phys. Rev. E **65**, 051923 (2002)
31. Manneville, P., Pomeau, Y.: Phys. Lett. **75A**, 1 (1979)
32. Bayés de Luna.: Clinical Electrocardiography: A Textbook. Futura (1998)
33. Brodde, O-E., Michel, M.C.: Adrenergic and muscarinic receptors in the human heart. Pharmacol. Rev. **51**, 651–689 (1999)
34. Morris, C., Lecar, H.: Voltage oscillations in the barnacle giant muscle fiber. Biophys. J. **35**, 193–213 (1981)
35. Holden, A.V., Biktashev, V.N.: Computational biology of propagation in excitable media models of cardiac tissue. Chaos, Solitons Fractals **13**, 1643–1658 (2002)
36. Hirsch, H., Huberman, B.A., Scalapino, D.J.: Theory of intermittency. Phys. Rev. A **25**, 519–532 (1982)
37. Cloherty, S.L., Dokos, S., Lovell, N.H.: A comparison of 1-D models of cardiac pacemaker heterogeneity. IEEE Trans. Biomed. Eng. **53**, 164–177 (2006)
38. Boyett, M., Honjo, M.R., Kodama, H.I.: The sinoatrial node, a heterogeneous pacemaker structure. Cardiovasc. Res. **47**, 658–687 (2000)
39. Nollo, G., Del Greco, M., Ravelli, F., Disertori, M.: Evidence of lowand high-frequency oscillations in human AV interval variability: evaluation with spectral analysis. Am. J. Physiol. **267**, Heart Circ. Physiol. **36**, H1410–H1418 (1994)
40. Demir, S.S., Clark, J.W., Giles, W.R.: Am. J. Physiol. Heart Circ. Physiol. **276**, H2221–H2244 (1999)
41. Kulka, A., Bode, M., Purwins, H.G.: On the influence of inhomogeneities in a reaction-diffusion system. Phys. Lett. A **203**, 33–39 (1995)
42. Ermentrout, G.B., Rinzel, J.: Reflected waves in an inhomogeneous excitable medium. SIAM J. Appl. Math. **56**, 1107–1128 (1996)
43. Żebrowski, J.J., Baranowski, R.: Type I intermittency in nonstationary systems—models and human heart rate variability. Phys. A **336**, 74–83 (2004)
44. Żebrowski, J.J., Baranowski, R.: Observations and modeling of deterministic properties of human heart rate variability. Pramana **64**, 543–562 (2005)
45. Gac, J.M., Żebrowski, J.J.: Nonstationary Pomeau-Manneville intermittency in systems with a periodic parameter change. Phys. Rev. E **73**, 066203 (2006)
46. Kuklik, P., Szumowski, L., Żebrowski, J.J., Walczak, F.: Physiol. Meas. **25**, 617–627 (2004)

Chapter 6
Nonlinear Systems of Fractional-Orders

6.1 Introduction

The recent increased interest in the study of dynamic systems of non-integer orders [1–4] originates from the prerequisite that most of the processes associated with complex systems exhibit non-local dynamics involving long time memory, and that fractional integral and fractional derivative operators share most of those characteristics. Many originally considered systems with lumped and/or distributed parameters can be more exactly described by fractional order systems [5] than by integer orders ones. In this field, specialists often use the term "fractional order (FO) calculus", or "fractional order dynamic system" where the adjective "fractional" actually means "noninteger" [1, 6]. Interest in the fractional calculus has had a significant impact on many fields of engineering, biology, chemistry and economics, resulting in a lot of research on the analysis and design of dynamic systems of fractional orders. The definitions most commonly used in the systems investigations are the Riemann–Liouville definition, the Grünwald-Letnikov definition, and the Caputo definition, as given below

$$ {}_aD_t^\alpha f(t) = \frac{1}{\Gamma(m-\alpha)}\left(\frac{d}{dt}\right)^m \left(\int_a^t \frac{f(\tau)}{(t-\tau)^{1-(m-\alpha)}}d\tau\right) \tag{6.1}$$

$$ {}_aD_x^\alpha[f(t)] = \lim_{h\to 0}\frac{1}{\Gamma(\alpha)h^\alpha}\sum_{k=0}^{\left[x-a/h\right]}\frac{\Gamma(\alpha+k)}{\Gamma(k+1)}f(t-kh) \tag{6.2}$$

$$ {}_aD_t^\alpha[f(t) = \frac{1}{(\alpha-n)}\int_a^t \frac{f^{(n)}(\tau)}{(t-\tau)^{\alpha-n+1}}d\tau \tag{6.3}$$

for $n-1<\alpha<n$ and where $\Gamma(\cdot)$ is the *Gamma* function, and [z] means the integer part of z.

Z. Trzaska, *Mixed Mode Oscillations (MMOs)*, Studies in Systems, Decision and Control 374, https://doi.org/10.1007/978-3-030-76867-6_6

The "memory" effect of these operators is demonstrated by (1) and (2), where the convolution integral in (1) and the infinite series in (2) reveal the unlimited memory of these operators, ideal for modeling hereditary and memory properties in physical systems and materials. The initial conditions for the fractional order differential equations with the Caputo derivative (3) are of the same form as the initial conditions for the integer-order differential equations [5]. In the sequel the approach is based on the fact that for a large class of functions, three above definitions are equivalent. Similarly to integer-order differentiation, fractional-order differentiation fulfils the relations

$$ {}_{a}D_{t}^{\alpha}({}_{a}D_{t}^{\beta}f(t)) = {}_{a}D_{t}^{\alpha+\beta}f(t) - \sum_{k=0}^{n}[{}_{a}D_{t}^{\beta-k}]_{t=a}\frac{(t-a)^{-\alpha-k}}{\Gamma(1-\alpha-k)} \tag{6.4} $$

$$ {}_{a}D_{t}^{\alpha}[\eta f(t) + \lambda g(t)] = \eta[{}_{a}D_{t}^{\alpha}f(t)] + \lambda[{}_{a}D_{t}^{\alpha}g(t)] \tag{6.5} $$

Moreover, fractional order systems may have other advantageous features that make them more suitable for the study of nonlinear dynamical systems when most of the desired depictions are not readily achieved by traditional models [7, 8].

More useful for the analysis of fractional-order systems, is given in the following state space form

$$ \begin{aligned} {}_{0}D_{t}^{q}x(t) &+ \varphi(x, q, \mathbf{A}, \beta, t) \\ &= Ax(t) + Bu(t), \text{ with } \varphi(x, q, \mathbf{A}, \beta, t) \text{ given for } t > \beta \\ y(t) &= Cx(t) + Hu(t) \end{aligned} \tag{6.6} $$

where $\boldsymbol{x}(t) \in \boldsymbol{R}^n, \boldsymbol{u}(t) \in \boldsymbol{R}^m, \boldsymbol{y}(t) \in \boldsymbol{R}^p$ are states, inputs, and outputs vectors of the system and $\boldsymbol{A} \in \boldsymbol{R}^{n\times n}, \boldsymbol{B} \in \boldsymbol{R}^{n\times m}, \boldsymbol{C} \in \boldsymbol{R}^{p\times n}, \boldsymbol{H} \in \boldsymbol{R}^{p\times m}$ denote constant element matrices, and $q < 1$ is the fractional commensurate order of the system. The fractional dynamic variables in the system of Eq. (6.6) are not states in the true sense of the classic 'state' space. Consequently, as the initialization function vector $\boldsymbol{\varphi}(\boldsymbol{x}, q, \boldsymbol{A}, \beta, t)$ is generally required, the set of elements of the vector $\boldsymbol{x}(t)$, evaluated at any point in time, does not specify the entire 'state' of the system. Thus, for fractional-order systems, the ability to predict the future response of a system requires the set of fractional differential equations along with their initialization functions, that is, complete Eq. (6.6). Note that by using the fractional differential equations, we get a total order of the system that is less than the number of differential equations.

In the alternative representation (6.6) for a linear time-invariant complex fractional system, the following state space model is convenient

$$ \begin{aligned} {}_{0}D_{t}^{q}\boldsymbol{x}(t) &= \boldsymbol{A}\boldsymbol{x}(t) + \boldsymbol{B}\boldsymbol{u}(t), \\ \boldsymbol{y}(t) &= \boldsymbol{C}\boldsymbol{x}(t) + \boldsymbol{H}\boldsymbol{u}(t), \end{aligned} \tag{6.7} $$

where $\boldsymbol{x} \in \boldsymbol{R}^n$, $\boldsymbol{u} \in \boldsymbol{R}^r$ and $\boldsymbol{y} \in \boldsymbol{R}^p$ are the state, input and output vectors of the system, and $\boldsymbol{A} \in \boldsymbol{R}^{n\times n}$, $\boldsymbol{B} \in \boldsymbol{R}^{n\times r}$, $\boldsymbol{C} \in \boldsymbol{R}^{p\times n}$, q is the fractional commensurate order. It has been shown that system is stable if the following condition is satisfied [9]:

$$|\arg(\mathrm{eig}(\boldsymbol{A}))| > q\pi/2 \tag{6.8}$$

where $0 < q < 1$ and $\mathrm{eig}(\boldsymbol{A})$ represents the eigenvalues of matrix $\boldsymbol{A}$.

Various concepts of stability analysis of fractional-order (FO) systems are available [9–11]. It has to be noted that in almost all cases the impulse responses of fractional order systems are related to the Mittag–Leffler function [7], which is effectively the fractional order analog of the exponential function, being common in the study of systems of integer orders. With this knowledge, it has been possible to better clarify the time responses associated with fractional order systems.

6.2 MMOs in a Fractional Order System

Good understanding of the mechanisms influencing the interaction between basic elements of each dynamic system has an essential meaning from the viewpoint of potential performance impacts. Identification of the complex processes, changing in the time, requires taking into account many different factors affecting the dynamics of the entire system. To determine the factors influencing on the effective functioning of the dynamic system and its variations, one can reach the significant help in the form of an effective research tool, namely, computer simulation. The knowledge of the possible dynamic behavior of fractional order systems (FOSs) is fundamental as most properties and conclusions of integer order systems (IOSs) cannot be simply extended to that of the FOSs. To demonstrate these facts we shall consider some selected cases of FOSs.

Let us begin with the fractional-order Volta's system [12], where integer-order derivatives are replaced by fractional-order ones and with $x(t)$, $y(t)$, and $z(t)$ standing as fractional state variables. Mathematical description of this system is expressed as

$$\begin{aligned} {}_0D_t^{q_1}x(t) &= -x(\mathrm{t}) - ay(t) - z(t)y(t), \\ {}_0D_t^{q_2}y(\mathrm{t}) &= -y(t) - bx(t) - x(t)z(t), \\ {}_0D_t^{q_3}z(t) &= cz(\mathrm{t}) + x(t)y(t) + 1, \end{aligned} \tag{6.9}$$

where q_1, q_2, and q_3 are the derivative orders. The total order of the commensurate system is $q = (q_1, q_2, q_3)$.

The vector representation of (6.9) is

$$D^q\boldsymbol{x} = f(\boldsymbol{x}) \tag{6.10}$$

where $\boldsymbol{q} = [q_1, q_2, q_3]^T$ for $0 < q_i < 1$, $(i = 1, 2, 3)$, $\boldsymbol{f} \in \boldsymbol{R}^3$ and $\mathbf{x} \in \boldsymbol{R}^3$. The stationary points of (6.10) are calculated via solving the following equation

$$\boldsymbol{f}(\boldsymbol{x}) = \mathbf{0} \tag{6.11}$$

and we suppose that $P^* = (x^*{}_1, x^*{}_2, x^*{}_3)$ is its stationary point.

Conditions for asymptotic stability of (6.10) result from the following statements. For a system of commensurate fractional orders the stationary points are asymptotically stable if all the eigenvalues λ_i $(i = 1, 2, 3)$ of the Jacobian matrix $\boldsymbol{J} = \partial \boldsymbol{f}/\partial \mathbf{x}$, evaluated at the stationary point P^* satisfy the condition (6.8) for all eigenvalues λ_i.

Due to the complex form of the established system of nonlinear equations of fractional orders, the determination of their analytical solution is not an easy task and therefore, to examine the effect of various system parameters on the periodic dynamical process, we can perform numerical calculations applying a computer program MATLAB with using effective numerical integration procedures. We solved (6.10) with the variable step ODE23 procedure from MATLAB with RelEr = AbsEr $= 10^{-8}$ and $0 \le t \le 50$ s and nonzero initial conditions.

Computed results for system parameter values $(a, b, c) = (0.3\ 0.5\ c)$ with c taken as bifurcation parameter and initial conditions fixed at $(x(0)\ y(0)\ z(0)) = (18\ 5\ 2)$ are shown in Fig. 6.1a. They represent the time varying fractional system variables in the steady state. The corresponding phase space portrait of the given system is shown in Fig. 6.1b.

Looking at the eigenvalues of the Jacobian it is easy to state that they correspond to a saddle node stationary point which involves instability of the system response for the wide range of possible values of the parameter c. Moreover, the above results can be used in many possible applications of FOSs and help assessing whether the latter are capable of addressing the industry's problems. As is well known, there exist some practical applications of MMOs in which the existence of non-overshooting step responses is critical, and consequently designing non-overshooting feedback

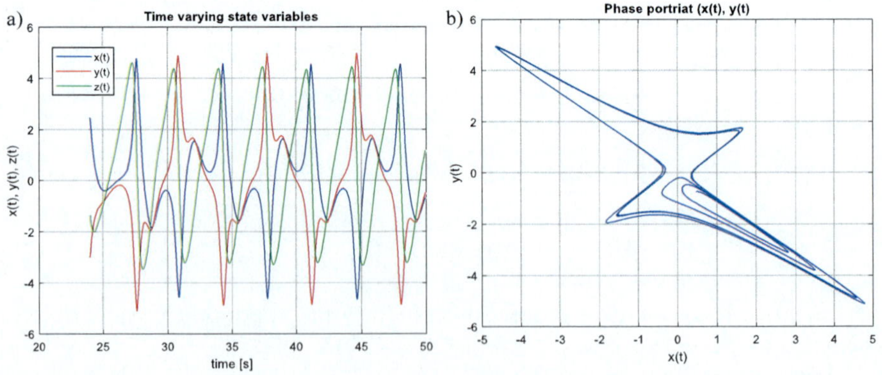

Fig. 6.1 Simulation results for Volta system of fractional orders $q_1 = q_2 = q_3 = 0.98$: **a** state variables: $x(t)$—in blue, $y(t)$—in red, $z(t)$—in green, **b** phase trajectory $y(t)$ versus $x(t)$

control systems is of great importance. Therefore, these conditions are beneficial in predicting the existence of an overshoot in the step response of a fractional-order model whose numerical simulation is not simple. The effectiveness of the given conditions was confirmed by some numerical examples.

The next problem provides a brief retrospective look on fractional calculus applications for studies of MMOs in nonlinear complex electric circuits. It is worth to emphasize, that the need to numerically estimate the state of an electric nonlinear circuit with fractional order derivatives and integrals arises frequently in many fields, especially in electronics, telecommunications, automatic control, power systems, and digital signal processing [6]. Systematic modeling methods and computer in MATLAB procedures are of great help tools that enable fractional calculus and its application to solving circuits of fractional order.

In this subsection the effects of fractional dynamics in nonlinear circuits are studied. In particular, Chua's standard circuit is modified to include fractional order elements. It is worth mentioning that fractional order Chua's circuit manifests to be an excellent paradigm for generation of a multitude of different dynamical phenomena and can thus obviate the need to consider many different models to simulate those phenomena. One of the main reasons behind Chua's circuits' popularity is their flexibility and generality for representing virtually many practical structures, including those undergoing dynamic changes of topology [13].

Let us consider the fractional order circuit shown in Fig. 6.2a which represents a modified Chua's circuit [14]. This circuit is different from the standard Chua's circuit in that the piecewise-linear nonlinearity is replaced by an appropriate cubic nonlinearity which leads to a very particular behavior. In this case, the circuit forms a nonlinear oscillator comprising a nonlinear resistor with a cubic characteristic $I_n(x_1)$, three fractors represented by an inductor and two supercapacitors, a current controlled current source I and a biasing constant current source a. Its comportment depends on all six constant parameters involved. It can exhibit a large spectrum of dynamical behaviors such as the relaxation, mixed-mode oscillation, bifurcation and

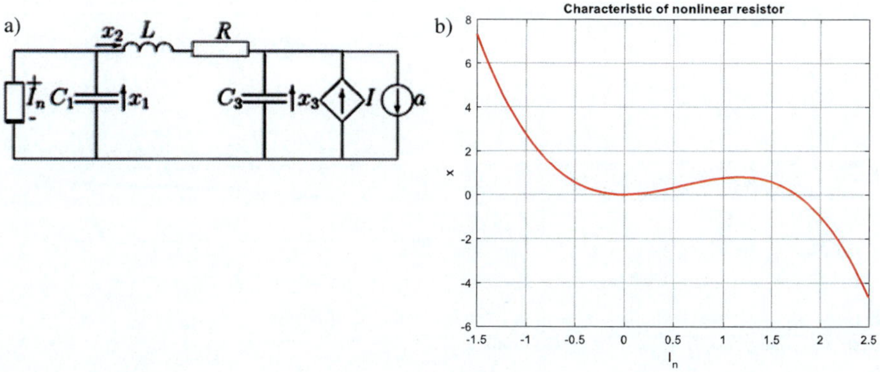

Fig. 6.2 Modified Chua's circuit: **a** scheme with fractors C_1, L, C_3 and $I = (1 + b)x_2$, $a =$ const > 0, $b =$ const > 0, $I_n = \alpha x_1^2 + \beta x_1^3$, $\alpha < 0$ and $\beta > 0$, **b** characteristic of nonlinear resistor

chaos. The mathematical description of dynamical components is based on general circuit laws and properties of the fractors and the studied circuit can be described as follows

$$C_1 \frac{d^{q_1} x_1}{dt^{q_1}} = -x_2 + \alpha x_1^2 + \beta x_1^3$$

$$L \frac{d^{q_2} x_2}{dt^{q_2}} = x_1 - x_3 - R x_2 \tag{6.12}$$

$$C_3 \frac{d^{q_3} x_3}{dt^{q_3}} = a - b x_2.$$

where q_1, q_3 and q_2 denote fractional orders of the supercapacitors C_1, C_3 and of the real coil L, respectively.

It is worth mentioning that even in the case of integer orders $q_1 = q_2 = q_3 = 1$ the above circuit can exhibit a number of exceptional behaviors depending on circuit linear element parameters, the form of the nonlinear resistor characteristic as well as on initial conditions. Of particular interest here are the mixed-mode oscillations (MMOs). For instance, assuming $C_1 = 0.01$F, $L = 1$H, $C3 = 1$F and a = 0.0005A, b = 0.0035, α = 1.5 and β = −1 with $q_1 = q_2 = q_3 = 0.9$ we get eigenvalues $\lambda_1 = 0$ and $\lambda_{2,3} \approx 1.0107809162 \pm 0.0153011315j$. Such stationary point is the unstable focus node. The MMOs with respect to the state variable $x_1(t)$ are presented in Fig. 6.3a. The simulations were performed without applying the short-memory principle for time step $h = 0.001$ with the following initial conditions: $x(0) = 0$, $y(0) = 0$, and $z(0) = 0$. Computer simulations also show the limit cycle (Fig. 6.3b) exhibiting MMOs in the studied circuit.

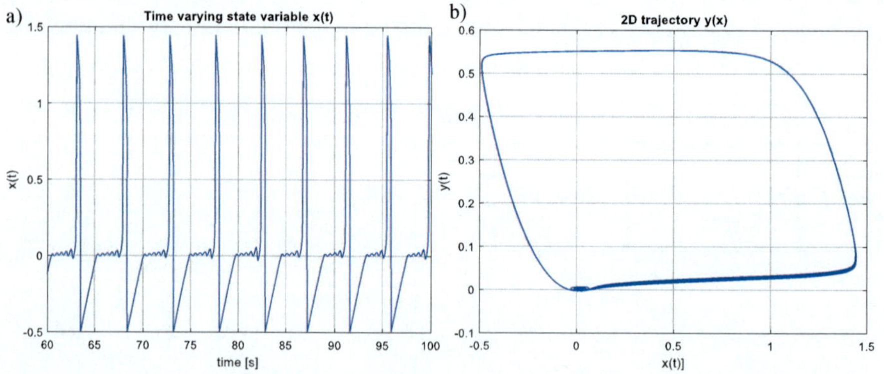

Fig. 6.3 Simulation results for a modified Chua's circuit of fractional orders $q_1 = q_2 = q_3 = 0.9$: **a** state variable $x_1(t)$, **b** trajectory $x_2(t)$ versus $x_1(t)$,

6.3 MMOs in Bipolar Pulsed Electroplating

Over the last two decades, electrochemical processes have become as versatile techniques for semiconductor and related microelectronic manufacturing. Recently, electro-deposition (ECD) of thin copper layers for multilevel interconnections in microelectronic devices has become largely accepted for mainstream silicon chip manufacturing, especially for high speed logic devices [1]. It is worth to emphasize, that much interest in ECD has evolved due to: (*i*) the low temperatures applied, (*ii*) the ability to coat geometrically complex or non line-of-sight surfaces of porous products, (*iii*) the capability to control the thickness, composition, and microstructure of the deposit, (*iv*) the possible improvement of the substrate/coating bond strength, and (*v*) the widespread availability and low cost of manufacturing equipment. Several applications of specific types of ECD processing systems, each with its own adequate requirements, met with great recognition amongst specialists in leading research fields and in top technological branches of industry. Depending on conditions of the process realization it is possible to produce thin layer materials with particular properties that influence importantly not only on research and modern fabrication intensifications but also on numerous domains of the top microelectronic technologies. Most of processes associated with complex ECD systems have non-local dynamics involving long in time memory, and fractional integral and fractional derivative operators exhibit some of those characteristics. Many practical ECD systems with lumped and/or distributed parameters can be accurately described in terms of fractional order models. Newly produced nano-devices need superior accuracies in component's dimensions and chances of correlating their hierarchies to models of the fractional order $0 < q < 1$, which lead to satisfied results when matching them with experimental data. Such models have a two-fold purpose: first, they can account all the facts discovered experimentally, and second, they can be able predict the system behavior under various conditions. The models allow us the exhibility to try and isolate from the matrix the contribution of a few key parameters in an attempt to predict the dominant effects as well as the variation of the process. It is well known, that the ECD process appears as a competition between nucleation and grain growth. In charging and discharging of a pulse, especially for short pulses, the double layer distorts the pulse current [15], and affects the over potential response acting on the electrolyte [12].

In the performed study the current–voltage relations were used to gain insights into the coupling mechanisms in relations charge–concentration of the solution being ultimately responsible for different pattern of output quantities and to identify the relationship between them. Various waveforms of the current supplying the ECD reactor are in practical use. However, presently all major process realizations concentrate around the PPRC (periodic pulse reverse current) electrodepositing technology and the general metal finishing products. Pulse current ECD is able to produce layers with more uniform particle distribution and better surface morphology than those obtained under direct current. This is common activity on rough parts or when a bright finish is required. Moreover, in certain processes, the management of electrolyte chemical concentrations is critical to ensure consistent results. During the thin layer metal ECD the complex reaction of charge transfer with intermediate adsorption takes place at least in two elementary steps: reagent adsorption is the first and product desorption is the second.

The ECD process can be described by a mathematical model derived from the Maxwell equations and mass balance equations as follows

$$\frac{d^{q_1}x_1}{dt^{q_1}} = -ax_1 + ax_3 + \alpha x_1^2 + \beta x_1^3$$
$$\frac{d^{q_2}x_2}{dt^{q_2}} = -bx_2 + cx_3 \tag{6.13}$$
$$\frac{d^{q_3}x_3}{dt^{q_3}} = dx_1 - dx_3 - fx_2 + fI_c$$

where state variable x_1 and x_3 denote potentials at the reactor input and at the processed detail, respectively. The state variable x_2 denotes the current in the electrolyte. The current supplying the reactor is marked as I_c. Values of the constant coefficients in Eqs. (6.13) depend on the construction of the reactor and properties of the electrolyte. Moreover, the fractional orders fulfill the relation $m-1 < q_k < m$, where m denotes a nearest integer and $k = 1, 2, 3$.

Figure 6.4 shows a non-linear relationship observed between the thin-layer diffusion current and the voltage applied to the double layer capacitance zone. It takes into account the Butler-Volmer model of electrode kinetics equation relating the surface over potential at the electrode–electrolyte interface to the current density [1]. Frequently it has a cubic form.

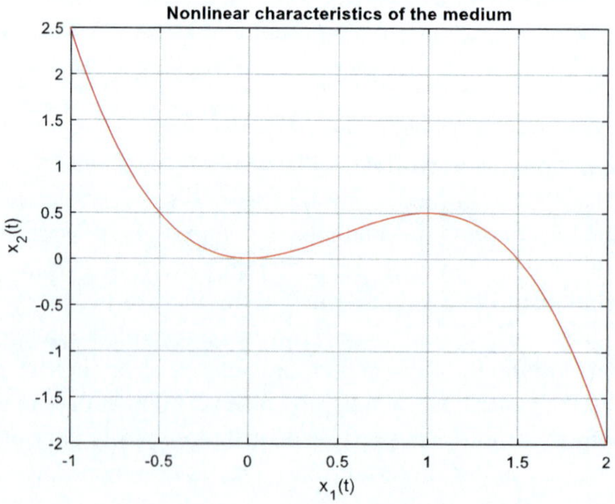

Fig. 6.4 Non-linear relationship between diffusion current x_2 (t) and voltage x_1(t) on the double layer

To determine the solution of (6.13) we have used a special numerical procedure based on the Bagley-Torvik scheme [16]. Taking into account parameters: $a = 0.00025$, $\alpha = 0.375$, $\beta = 0.25$, $b = 0.0002$, $c = 0.2$, $d = 0.001$, $f = 1.05$, and $I_c = -110$ for $0 < t < T_1$ and $I_c = 110$ for $T\text{-}T_1 < t < T$ with $I_c(t) = I_c(t + T)$ and applying the mentioned numerical procedure implemented in Matlab we get the solutions shown in Figs. 6.5 and 6.6. In all performed simulations the fractal orders were equal to $q_1 = q_2 = q_3 = 0.9$. The obtained results indicate on important influences of the period of the bipolar pulse current supplying the reactor. Moreover it is worth to mention that the rate of the period fulfilling with constant magnitude of the supplying current causes important changes in the output signals of the reactor.

Electrodepositions of such materials as copper surface thin-layers and graphene flakes with micro- and nano-crystalline structures can be successfully modeled by fractional order differential equations rather than traditional integer order ones. The dimension of the fractional order sets used to model these structures is of primary importance as it determines the scaling and ultimately the sizing effects for these particular materials. Contribution of fractional order capacities at the reactor input is much higher than the contribution of the external current source to the instantaneous

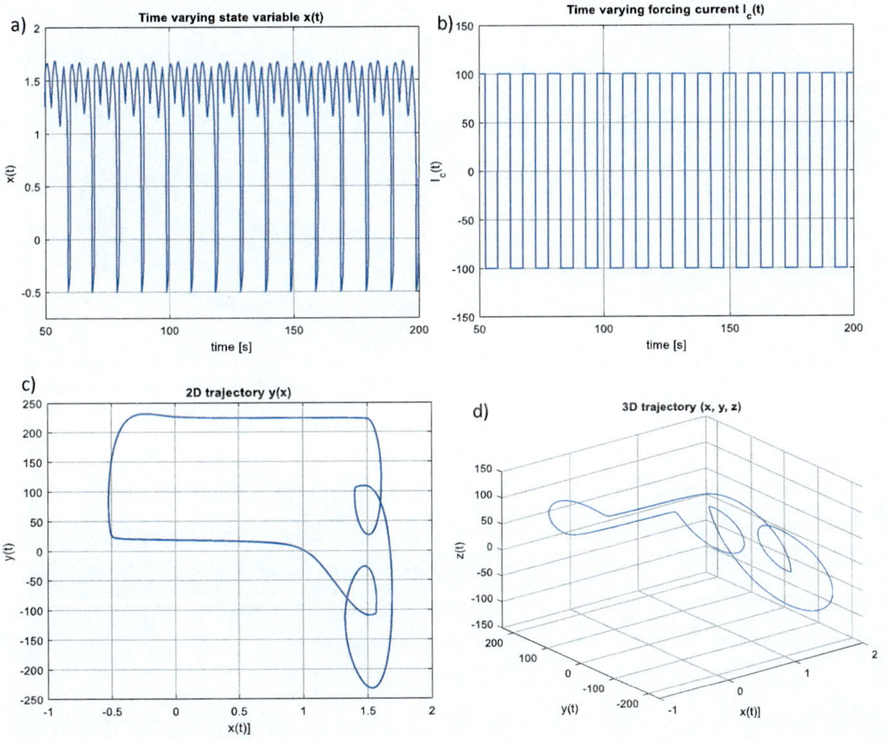

Fig. 6.5 Simulation results: **a** time varying potential at the reactor input, **b** supplying current, **c** 2D phase portrait, **d** 3D phase portrait

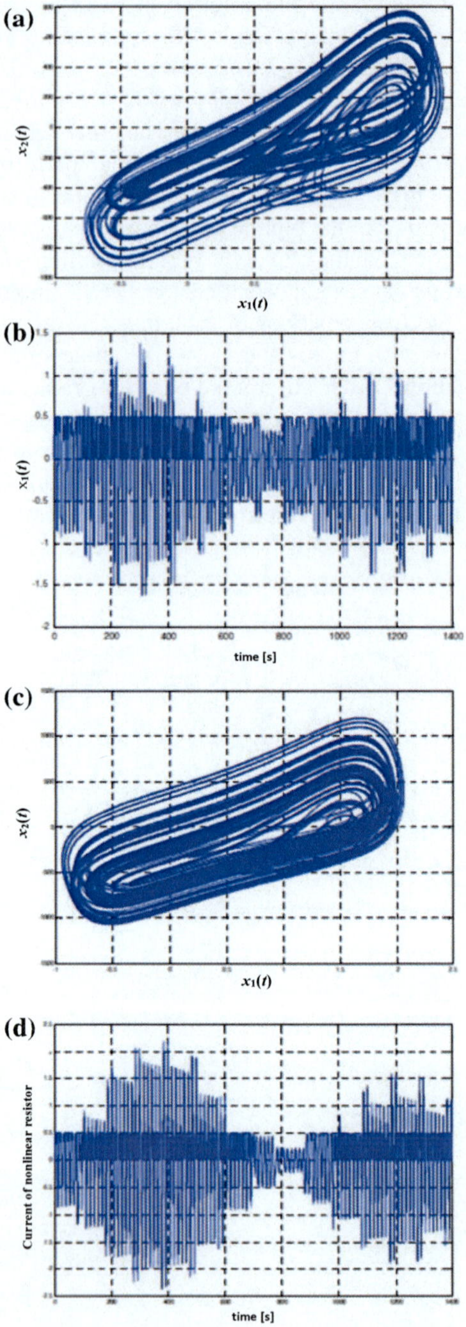

Fig. 6.6 Phase trajectories and currents in the diffusion layer versus time for: **a**, **b** T = 4T1 = 100s; **c**, **d** T= 4/3T1 = 100s

behavior of the whole system. However for a given reactor fulfilled by a strictly composed electrolyte the period of the bipolar pulsed current influences importantly on the instantaneous state variables in the systems and this phenomenon must be taken into account for effective control of manufacturing processes of the reactor.

References

1. Podlubny, I.: Fractional Differential Equations. Academic Press, New York (1999)
2. Trzaska, Z.: Fractional-order systems: their properties and applications. Elektronika **10**, 137–144 (2008)
3. Trzaska, Z.: Fractional order model of Wien Bridge oscillators containing CPEs. In: Proceedings of MATHMOD'09 Conference, Vienna, pp. 357–361 (2009)
4. Trzaska, Z.: Chaos in fractional order circuits. Electr. Rev. **86**(1), 109–111 (2010)
5. Li, C.P., Zhang, F.R.: A survey on the stability of fractional differential equations. Eur. Phys. J. Spec. Top. **193**(1), 27–47 (2011)
6. Petras, I.: Fractional-Order Nonlinear Systems Modeling, Analysis and Simulation. Springer, Heidelberg, Dordrecht, London, New York (2011)
7. Li, Y., Chen, Y., Podlubny, I.: Mittag-Leffler stability of fractional order nonlinear dynamic systems. Automatica **45**(8), 1965–1969 (2009)
8. Petras, I.: Stability of fractional-order systems with rational orders: a survey. Fract. Calculus Appl. Anal. **12**(3), 269–278 (2009)
9. Caponetto, R., Dongola, G., Petras, F.I.: Fractional Order Systems. Modeling and Control Applications. World Scientific, London, Singapore (2010)
10. Han, X., Bi, Q., Zhang, Ch., Yu, Y.: Study of mixed-mode oscillations in a parametrically excited van der Pol system. Nonlinear Dyn. **77**, 1285–1296 (2014)
11. Han, X., Xia, F., Zhang, Ch., Yu, Y.: Origin of mixed-mode oscillations through speed escape of attractors in a Rayleigh equation with multiple-frequency excitations. Nonlinear Dyn. **88**, 2693–2703 (2017)
12. Liu, Z., West, A.C.: Modeling of galvanostatic pulse and pulsed reverse electroplating of gold. Electrochim. Acta **56**, 3328–3333 (2011)
13. Chua, L.O.: http://www.scholarpedia.org/article/Chuacircuit
14. Marszalek, W., Trzaska, Z.: Mixed-mode oscillations in a modified Chua's circuit. Circ. Syst. Sig. Process **29**, 1075–1087 (2010)
15. Vakarin, E.V., Badiali, J.P.: Electrochemical response of flexible disordered host matrices under different insertion conditions. Electrochim. Acta **56**, 3526–3529 (2011)
16. Datta, M.: 'Electrochemical processing technologies in chip fabrication: challenges and opportunities.' Electrochim. Acta **48**, 2975–2985 (2003)
17. Josell, D., Wheeler, D., Moffat, T.P.: 'Superconformal electrodeposition in Vias.' Electrochem. Solid-State Lett. **5**(4), C49–C52 (2002)
18. Kim, B., Ritzdorf, T.: 'Electrochemically deposited Tin–Silver– Copper ternary Solder alloys.' J. Electrochem. Soc. **150**(2), C53–C60 (2003)
19. Hu, F., Chan, K.C.: Equivalent circuit modelling of Ni–SiC electrodeposition under ramp-up and ramp-down waveforms. Mater. Chem. Phys. **99**, 424–430 (2006)
20. Maxwell, J.C.: A Treatise on Electricity and Magnetism, 3rd edn., vol. 2. Clarendon, Oxford, pp. 68–73 (1892)
21. Goonetilleke, P., Roy, D.: Voltage pulse-modulated electrochemical removal ofcopper surface layers using citric acid as a complexing agent. Mater. Lett. **61**, 380–383 (2007)
22. Manens, A., Miller, P., Kollata, E., Duboust, A.: Advanced Process ControlExtends ECMP Process Consistency. Solid-State Technology (2006)
23. Trzaska, M., Trzaska, Z.: Electrochemical Impedance Spectroscopy in Materials Science. Publishing Office of the Warsaw University of Technology, Warsaw (2010)

24. Trzaska, M., Trzaska, Z.: Energetic process modelling of thin-layer electrocrystallization. Electr. Rev. **87**(5), 173–175 (2011)
25. Trzaska, M., Trzaska, Z.: Straightforward energetic approach to studies of the corrosion performance of nanocopperthin-layers coatings. J. Appl. Electrochem. **37**, 1009–1014 (2007)
26. Tavazoei, M.S., Haeri, M.: A note on the stability of fractional order systems. Math. Comput. Simul. **79**(5), 1566–1576 (2009)
27. Trzaska, Z.: Impact oscillators: fundamentals and applications. J. KONES Powertrain Transp. **22**(1), 307–314 (2015)
28. Litak, G., Margielewicz, J., Gaska, D., Yurchenko, D., Dabek, K.: Dynamic response of the spherical pendulum subjected to horizontal Lissajous excitation. Nonlinear Dyn. **102**, 2125–2142 (2020)
29. Liu, N., Ouyang, H.: Friction-induced vibration considering multiple types of nonlinearities. Nonlinear Dyn. **102**, 2057–2075 (2020)
30. Lozi, R., Abdelouahab, M.-S., Chen, G.: Mixed-Mode Oscillations Based on Complex Canard Explosion in a Fractional-Order Fitzhugh-Nagumo Model. ffhal-02541322f (2020)
31. Trzaska, Z.: Matlab solutions of chaotic fractional order circuits. In: Saleh, A. (ed), Engineering Education and Researche Using MATLAB, Chapter 10, InTech, Rijeka (2014)
32. Qian, Y., Meng, W.: Mixed-Mode Oscillation in a Class of Delayed Feedback System and Multistability Dynamic Response. Hindawi, Complexity, vol. 2020, Article ID 4871068, 18 pages (2020)
33. Chena, Z., Chen, F.: Mixed mode oscillations induced by bi-stability and fractal basins in the FGP plate under slow parametric and resonant external excitations. Chaos Solitons Fraktals **137**, 109814 (2020)

Chapter 7
MMOs in Mechanical and Transport Systems

7.1 MMOs in Nano-oscillators

7.1.1 Introduction

Over the past several decades, scientific and engineering progress has focused on improving the manufacturing and applications of nanoscale final products, which are now becoming a key factor in society's security, economic well-being and quality of life improvement. The tasks that are assigned to the materials and end products currently used in practice are matching increasingly important in terms of their better adaptation to special technological features and reduction of their wear and tear. Presently, nanotechnology is focusing on the formation of purposeful materials, devices, and systems through the management of matter on the nanometer scale and the utilization of new occurrences and attributes at that scale. Nanometric structure materials have notably advantageous properties and constitute an effective alternative to conventional materials. The ability to control the structure of the particular device area, which presents a great potential for use in the development of new material properties can be reached by production process with the electrocrystallization method stemming from reactions of oxidation and electrochemical reduction.

The main objective of this chapter is to bring together a unified physical basis and mathematical tools for analyses of mixed mode oscillations under excitations essentially unharmonic, non-smooth or may be discontinuous shapes in time. The incorporation of nonlinearity in the performed analysis is one of the main features of this study what requires special attention both in modeling and solution procedures. It is known that possible transitions to nonsmooth limits can make investigations especially difficult. This is because the dynamic methods were originally developed within the paradigm of smooth motions based on the classical theory of differential equations. However, up to-date, many theoretical and applied sciences concentrate on high-energy phenomena accompanied by strongly non-linear spatio-temporal behaviors making the classical smooth methods inefficient in many cases. For instance,

Z. Trzaska, *Mixed Mode Oscillations (MMOs)*, Studies in Systems, Decision and Control 374, https://doi.org/10.1007/978-3-030-76867-6_7

such phenomena occur when dealing with dynamical systems under constraint conditions, friction-induced vibrations, structural damages due to cracks, liquid sloshing impacts, and numerous other problems of nonlinear physics. Possible approaches to such problems can be constructed by generating system models exhibiting essentially nonlinear/anharmonic behaviors as their intrinsic properties. The main attention is focused on properties of dynamical systems containing electromechanical components made with aids of nanomaterials like graphene and carbon nanotubes possessing a wide variety of outstanding properties.

Graphene, an atomically thin layer of carbon atoms tightly packed into a two-dimensional (2D) honeycomb lattice (Fig. 7.1a) has extremely high stiffness along the fundamental plane and it is the strongest material ever measured experimentally [1, 2]. The length of the carbon–carbon bond in graphene is approx. d = 0.142 nm. The graphene slices are arranged in piles at intervals of approx. 0.335 nm, form a graphite. This means that a stack of 3 million graphene layers would be only one millimeter thick. The separation of graphite layers was at the heart of the discovery of graphene [3]. Relative experiments have shown that Young's modulus of graphene is about 1.0 TPa [2]. Graphene, as a relatively recently discovered nanomaterial and revealing interesting peculiar electronic and mechanical properties, appears to be a promising candidate for future wide applications in electronics, mechanics, cryptology, control engineering and in materials science. The superior mechanical properties of graphene make it ideally suited for nanoresonators, since Young's modulus is a key factor to improve the frequencies of nanoresonators [3, 4]. It can achieve high resonant frequencies that can be externally tuned over a wide range with moderate applied voltages. Moreover, when a graphene membrane is configured as a suspended vibrating field-effect device then its charge-tunable conductance and large electrical mobility allow efficient transduction of mechanical vibration [4–7].

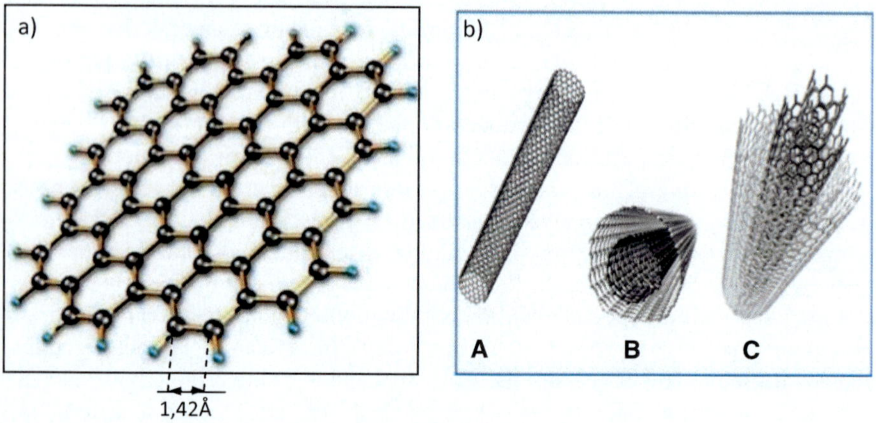

Fig. 7.1 Carbon nanostructures: **a** graphene (Grp), **b** carbon nanotubes (CNTs): A—single-walled, B—double-walled, C—multi-walled

Another of the allotropic configurations of carbon are Carbon Nano Tubes (CNTs). They are formed by folding graphene layers into cylindrical single- or multi-walled objects, which are hollow inside, with a diameter of 1.2–2.0 nm in the case of single-walled nanotubes and up to over 25 nm for multi-wall nanotubes and a length of approx. 1 μm or longer (Fig. 7.1b) [5]. Their basic structure is the hexagonal graphene network with the densest packing of carbon atoms. The unique and diverse structure of the graphene planes resulting from the method of folding graphene planes determines their unique properties and the large length-to-diameter ratio allows them to be treated as one-dimensional (1D) elements. The walls of adjacent tubes in multi-walled CNTs are held together through the van der Waals forces, which are the inter-atomic weak forces between nonbonded atom pairs. The van der Waals reciprocal actions between multi-walled CNTs allow them to chute and turne easily relatively to each other. CNTs are very versatile components in nanoelectromechanical systems (NEMS) due to their excellent mechanical and electrical properties that allow high tunability by an external electric field [1–3]. NEMS devices based on CNTs [4–7] provide a unique platform to test mechanics at the nanoscale. They boost the limits of detection to higher precision for force [8] and mass sensing [9]. Due to the small diameter of the CNTs, they can easily be brought into a non-linear oscillation state [10]. Moreover, it has been shown [11] that the nonlinear dynamics of CNTs can be tuned over a wide range of oscillation frequencies, which makes NEMS nanotubes ideal candidates for the wide implementation of sensors based on nonlinear characteristics and for the study of fundamental not only physical but also biological phenomena [12, 13]. A carbon nanotube behaves like a semi-flexible polymer in the sense that it can bend and stretch to great extents [14]. In consequence, nonlinear characteristics in nanotube resonators are important and result in an large variety of various dynamics [15–17].

In NEMS based on carbon nanotubes and graphene, damping is found to be strongly dependent on the amplitude of vibrations, and the damping force appears nonlinear in nature. For this reason, nonlinear damping can have a significant impact on the dynamics of micromechanical systems and can improve the benefit aspect in both nanotube and graphene resonator applications. Bottom-up fabricated nanoresonators based on nanomaterials like graphene and carbon nanotubes possess a wide variety of outstanding properties. Moreover, coupled nano- and micro-mechanical resonators attract continuously considerable attention of researchers and practitioners. Their mechanical vibrations can give rise to rich linear and nonlinear dynamics such as nonlinear mode coupling [12–20], oscillation localization [21, 22], state synchronization [23], mixed mode oscillations and chaos [24], parametric mode splitting and coherent phonon manipulation [25]. In a system with the quantum state [16], nonlinear mode coupling appears as a way to perform quantum non-destructive measurements [26].

7.1.2 Nanoelectromechanical Oscillators

Oscillators producing continuous periodic signals from direct current supplying devices are central to modern communications systems, electrical actuation and detection, medical diagnosis and treatments as well as miscellaneous applications including timing references and frequency modulators [1–7]. Up to-date, many theoretical and applied sciences concentrate on high-energy phenomena accompanied by strongly non-linear spatio-temporal behaviors making the classical smooth methods inefficient in many cases. For instance, such phenomena occur when dealing with dynamical systems under constraint conditions, with friction-induced vibrations, structural damages due to cracks, liquid sloshing impacts, and numerous other problems of nonlinear physics and engineering. However, possible approaches to such problems can be inspired by generating corresponding system models exhibiting essentially nonlinear/anharmonic behaviors as their intrinsic properties.

To confirm this fact, we use a graphene oscillator as an active element to generate a frequency mixed mode signal and obtain an efficient audio signal transmission. In these oscillators, the self-sustaining MMOs are generated and transferred to a load at room temperature using simple electrical circuitry. The prototype of a graphene voltage-controlled oscillator is presented in Fig. 7.2a. It consists of suspended strip of chemical vapour deposited (CVD) graphene, metal electrodes and a clamping structure made from epoxy that defines a rectangulr graphene layer in dimensions 2–4 mm. The graphene is suspended over a metal (Cu) local gate electrode on an insulating substrate.

The oscillator can achieve high resonant frequencies while having the mechanical compliance needed for tunability, and only require a small on-chip area. The vibration of the graphene is driven by applying a direct current (d.c.) and radio frequency excitation to the gate, and read out by applying a second d.c. bias to the drain. On resonance, the vibration of the graphene modulates the charge density, which in turn modulates the conductance and drain current. The large electronic mobility of

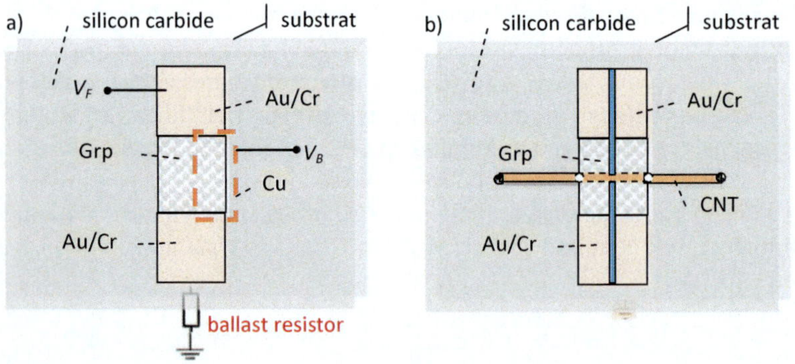

Fig. 7.2 Graphene voltage-controlled oscillator: **a** basic structure, **b** structure with additional electrode CNT

graphene, combined with its high mechanical compliance (spring constant ranging from 0.1 to 1 Nm), leads to efficient electromechanical coupling. The small gate spacing (200 nm, equivalent to a static capacitance of 44 μF/mm^2) and large sample size also contribute to a high Spectral Band Replication (SBR), even at room temperature [20], facilitating direct electrical transduction of the mechanical motion. The Au/Cr electrodes are colored yellow, the graphene Grp plate is colored gray, and the local gate electrode (Cu) is marked by dashed red rectangle. To operate the oscillators, periodic forces of electric origin actuate mechanical vibrations of the graphene sheet. The sheet is free to carry out vertical oscillations and is fastened to an insulating substrate. Between the graphene sheet and bulk electrodes there exists an oxide layer creating a resistive link, through which electrons can be transfered. Moreover, below the structure is placed a gate electrode which can adjust the strength of electrical field acting on the movable part of the graphene sheet. Therefore, the graphene sheet can be viewed as an integrated element with the gate and bulk electrodes. When the graphene sheet is under the potential of external source the electrical field from the gate exerts an unbalanced force on the movable part of the sheet and it deflects. The deflection changes in its turn the electrical potential of the sheet, and thereby the external source charge flow to the sheet. This means that a feedback of the mechanical motion is created but gate electrode is at a larger distance from the graphene sheet and thus the capacitive coupling between the sheet and the gate can be ignored.

7.1.3 State Variable Equations of Nanoelectromechanical Oscillators

Recent progress in the fabrication of graphene products leads now to possibilities of successfully exploring the damping processes in systems with components of one or more atomic-scale dimensions. If the nanooscillator vibrates with large amplitudes away from the equilibrium point, then its nonlinear dynamics cannot be neglected and the vibration takes an anharmonic form. In general, the damping observed in these systems can be expressed by a nonlinear damping force. To show that nonlinear damping in a graphene nano oscillator is a robust phenomenon, we study an electromechanical resonator with a graphene sheet under tensile stress (schematic of Fig. 7.2a). Nonlinear dissipation of energy is investigated as one possible attenuation mechanism of the unstable mechanical vibration.

Recall that to operate the oscillators, periodic forces of electric origin are applied to actuate mechanical vibrations. In addition, the graphene sheet is free to perform vertical oscillations and is clamped to an insulating substrate.

One useful way to describe the motion of the oscillator is by its state variable equation. If the oscillator is actuated to larger amplitudes away from the equilibrium point, nonlinear dynamics cannot be neglected and the resulting oscillation will be anharmonic.

Let $v_F(t) = V_m\cos(\omega t)$ be a harmonic excitation considered as a control quantity. Denote by x_1(t) and $x_2(t)$ the deflection and velocity of the graphene sheet from the free position with respect to epoxy base, by m and b the effective mass of the graphene sheet and the damping factor, respectively. By applying the Euler–Bernoulli equation for vibrating clamped beams [15] the mass-damper-stiffness dynamic model for the considered graphene oscillator takes the following form

$$\dot{x}_1 = x_2$$

$$m\,\dot{x}_2 = -(b + cx_1^2)x_2 - \frac{\partial V}{\partial x_1} + F_{drive}\cos(\omega t) \tag{7.1}$$

where point stand for the time derivative and $V = V(x_1)$ is the potential energy. The term F_{drive} denotes the magnitude of the electro-mechanic driving force changing with angular frequency ω. It comes from the coactions of applied field from external source and constant field excited by the gate potential. The result is slowly oscillating terms at the time-scale of x_1, which carries the electromechanical feedback. The term F_{drive} is the interesting one that can give rise to the nonresonant instability mechanism. The formula determining this parameter takes the form

$$F_{drive} = \frac{CV^2}{4d^2} \tag{7.2}$$

where C is the capacity of the graphene sheet and the gate electrode, V is the magnitude of the supplying voltage and d is the distance of the sheet from the gate electrode at rest state.

From a dynamics viewpoint, it is convenient to consider energy components of the nano oscillating system. Usually, the expression for the potential energy is given by

$$\mathrm{V}(\mathrm{x}_1) = \frac{1}{2}\mathrm{a}_1\mathrm{x}_1^2 + \frac{1}{12}\mathrm{a}_3\mathrm{x}_1^4 + \frac{1}{30}\mathrm{a}_5\mathrm{x}_1^6 + \frac{1}{64}\mathrm{a}_7\mathrm{x}_1^8 \tag{7.3}$$

The coefficients a_k (k = 1, 3, 5, 7) are constant and their values depend on the fastening of the graphene sheet. It should be underlined that Eqs. (7.1) and (7.2) give a general form of nonlinear nanoresonators description, in the sense that, in the case of weak damping and weak anharmonicity, additional terms of second and third order (x_2^2,$x_1^2x_2^2$,x_2^3,$x_1x_2^3\cdot x_1^2x_2^3$,) must be taken into account what only leads to a renormalization of b and c.

In order to present the essential properties of such a system, it is most convenient to examine first the behavior of a relatively simple autonomous system in the form of an anharmonic oscillator with a cubic nonlinearity. Thus, a pre-deflected, non-vibrating graphene sheet free from any external excitation is examined firstly. It is described by the following state variable equation

$$\dot{x}_1 = x_2$$
$$\dot{x}_2 = -\alpha x_1 - \beta x_1^3 - \gamma x_2 - \delta x_1^2 x_2 \tag{7.4}$$

where $\alpha = a_1/m$, $\beta = -\frac{1}{3} a_3/m$, $\gamma = b/m$, $\delta = c/m$. Under such conditions, the dynamics of the system depends on the parameters of the graphene sheet, namely, mass, damping factor and degree of deflection. The system (7.4) presents a cubic anharmonic oscillator with nonstandard Hamiltonian nature [26], which can be viewed as a conservative nonlinear oscillator perturbed by the nonlinear damping component $\delta x_1^2 \dot{x_1}$. It is worth underline that in reason of the nonlinear damping term, Eq. (7.4) exhibits PT–symmetric nature what means that we have $x = -x$ and $t = -t$. Note also that the model (7.4) has three equilibrium points which are

$$P_0 = (0,\ 0); \quad P_{1,2} = \left(\pm\sqrt{\frac{-\alpha}{\beta}}\ ,\ 0\right); \tag{7.5}$$

So that the following considerations are more detailed, we set the parameter values as $\alpha = 0.5$ and $\beta = 0.5$ and have $P_0 = (0, 0)$ and $P_{1,2} = (\pm 1\ ,\ 0)$. This means that the system has one saddle equilibrium point at (0, 0) that is an unstable fixed point and two centers at $(\pm 1, 0)$.

All three equilibrium points are depicted in Fig. 7.3a, in which the two centers are represented as open triangles and the saddle point is represented by a square. The same diagram shows the oscillator trajectories in the case of negligible damping in the system, which corresponds to the assumption in Eq. (7.4) of parameters $\gamma = 0$ and $\delta = 0$. When we choose initial conditions in the local regions of either the positive center point or the negative one, the trajectories move in a clockwise direction and pass within single region of the space plane, with small-amplitude periodic oscillations. On the other hand, for initial conditions selected away from the points of equilibrium, the system exhibits periodic oscillations passing within the region of the phase plane containing both center points. The orbit that separates

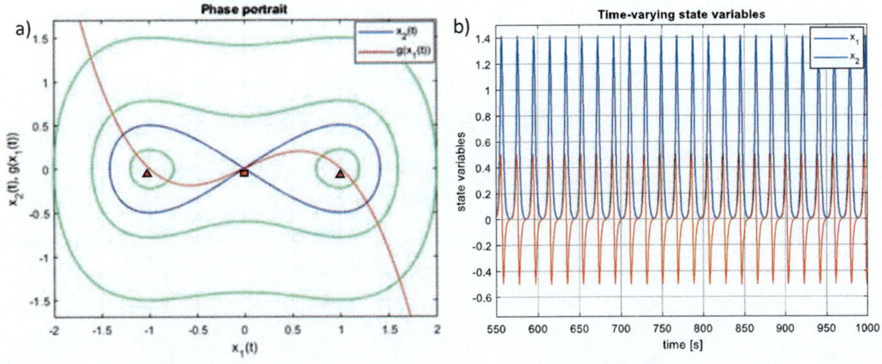

Fig. 7.3 Autonomous nanooscillator: **a** phase plane trajectories with $g(x_1(t)) = \alpha x_1(t) + \beta\, x_1^3(t)$, **b** time-varying state variables corresponding to homoclinic orbit

the single-center point and double-center point oscillatory regions in the phase space is known as homoclinic orbit, which is plotted in Fig. 7.3a with a pink line. Time evolutions of the steady state variables in the system (7.4) with $\delta = 0$ are presented in Fig. 7.3b. They correspond to initial conditions for the homoclinic trajectory taken from the state space region containing stationary point P_1. Note, that this gives the self-sustained oscillations.

To improve the performances of the nano oscillator within a wide frequency range outperforming linear resonant devices the adding of nonlinearities in its damping may play a vital role to solve such issues. Thus, having a nonlinear damping term of the form $\delta\ x_1^2\ x_2$ in the anharmonic oscillator Eq. (7.4) and investigating the system dynamics with respect to the nonlinear damping parameter, it is possible to identify significant differences in the waveform of resulting oscillations with respect to the previous case. First of all, due to the inclusion of nonlinear damping component, the symmetry of the system is annihilated instantly and the system has a parity and time-reversal (PT)—symmetry; this alters the stability of the equilibrium points. Further, for the positive values of δ, the equilibrium point in the region of the negative value of the first state variable becomes a source and repels nearby trajectories. The repelled trajectories are attracted by the fixed point in the positive value of the first state variable that acts as a sink. Therefore, the system has a dissipative nature in some regions of the phase space in which the trajectories are damped and attracted to the right region of the state space. At the same time, in other areas of phase space the system exhibits non-dissipative dynamics where the trajectories take the form of closed loops demonstrating periodic non-attenuated oscillations. The system is dissipative or non-distorting, depending on the location of the trajectory in the phase space according to the initial conditions. For fixed values of the system parameters $a = 0.5$ and $\beta = 0.5$ and for $F = 0$ the system is characterized by three stationary points: a stable focus at $(x, y) = (1, 0)$, a saddle point at (0,0) and an unstable focus at $(-1,0)$, which are clearly depicted in Fig. 7.4a as a filled triangle, square, and circle, respectively. Note that the stable focus point at (1,0) is linearly stable and nonlinearly

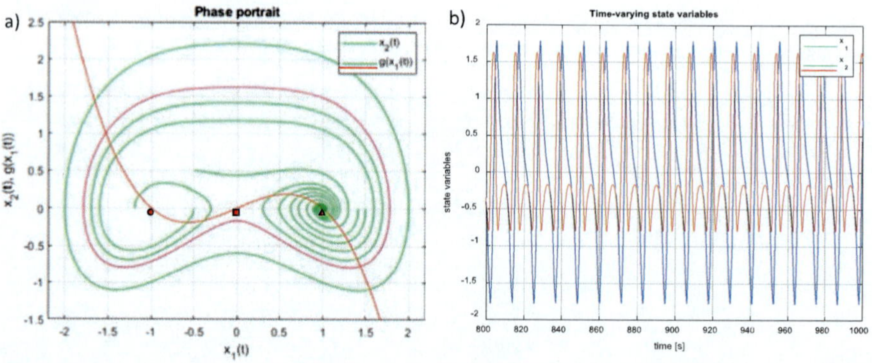

Fig. 7.4 Neutrally stable periodic orbits: **a** trajectory and nullcline in the phase space, **b** oscillations of x_1 and x_2

unstable. That is, the trajectories dissipate and approach to the stable focus only if we choose the initial conditions within some region in the phase-space. Otherwise, the orbits exhibit limit cycle oscillations. Hence, based on the initial conditions the system has either a dissipative or conservative nature. The homoclinic orbit is plotted in Fig. 7.4a with a pink line.

The underlying mechanism can be explained in terms of the total energy of the system as follows [27]. The total energy of the oscillator described by (7.4) can be expressed as

$$E = \frac{1}{2}\left[x_2^2 + f(x_1)\left(\delta x_2 - \frac{1}{2}\beta f(x_1)\right]\exp\left(\frac{\delta}{\sigma}\tan^{-1}\left(\frac{\delta x_1}{2\sigma x_2}\right)\right) + \frac{\alpha^2}{4\beta}\exp\left(\frac{\delta\pi}{2\sigma}\right)\right) \tag{7.6}$$

where $f(x_1) = x_1^2 + \frac{\alpha}{\beta}$ and $\sigma = \sqrt{2\beta - (\frac{\delta}{2})^2}$.

By substituting the initial conditions for the state variables (x_1, x_2), it is easy to establish that for some initial conditions the energy E is negative and for other ones it is positive. Therefore the system exhibits dissipative dynamics if $E < 0$, and when $E > 0$ it displays a conservative nature. In particular, inside the region of Fig. 7.4a limited by the homoclinic trajectory, the total energy of the system takes negative values and so the trajectories originated from this area have a dissipative nature and converge towards the stable focus when time varies.

On the other hand, initial conditions chosen with the outside of the homoclinic trajectory lead to positive values of E and hence the trajectories starting in this region exhibit conservative nature with neutrally stable periodic orbits what is visible in Fig. 7.4a.

Hence, the inclusion of nonlinear damping makes the system asymmetric and the unstable focus in the left potential region does not attract system trajectories.

The influence of nonlinear damping on the dynamics of graphene oscillators is quite well manifested in systems excited by a direct external periodic forcing. Now, the nonlinear damping term in Eq. (7.4) with $\delta > 0$ and $F_{drive} \neq 0$ can act as an energy-dissipating component as well as an energy-supplying component, which can give rise to systemic self-oscillations. Recall that the oscillation of a nanomechanical system is considered to be nonresonant if the excitation process is independent of the relation between the applied driving frequency and the self-oscillation frequency. Moreover, if the nano oscillator is actuated to larger amplitudes away from the equilibrium point, nonlinear dynamics cannot be neglected and the motion will be anharmonic. In the following, we assume that the external electrical field is polarized perpendicularly to the graphene sheet and is spatially uniform. In addition, the dynamics of the sheet is assumed to be homogeneous in the horizontal-direction and the graphene sheet is free to perform vertical oscillations.

As a demonstration, we use a graphene oscillator as the active element for mixed mode signal generation and achieve efficient audio signal output. Considering the above, the dynamics of an anharmonic oscillator with cubic nonlinearity in the presence of linear damping, nonlinear damping and periodic external forcing is investigated. The main goal is to determine the role played by non-linear damping in the development of MMOs and strategies for controlling such oscillations.

With the inclusion of an external periodic excitation of the form $F_{drive}\sin(\omega t)$, Eq. (7.4) can be rewritten as

$$\begin{aligned} \dot{x}_1 &= x_2 \\ \dot{x}_2 &= \alpha x_1 - \beta x_1^3 - \delta x_1 x_2 + F_{drive}\sin(\omega t) \end{aligned} \tag{7.7}$$

where F_{drive} and ω are the amplitude and angular frequency of the sinusoidal excitation, respectively. Now, in accord to Eqs. (7.7) and $F_{\text{drive}} > 0$ the symmetry of the equilibrium points is destroyed and in consequence they are moving along the x_1 axis with respect to the forcing amplitude F_{drive}. Following this effect, the dissipative and non-dissipative regions of the system phase space are also oscillating in time. Moreover, the area for dissipative vibrations is enlarged significantly. When the initial conditions take the zero values then the oscillation of the fixed points is governed by the stretching and folding actions, which ultimately can lead the system to mixed mode behavior. In the performed simulations, the values of oscillator parameters are set at $\beta = 0.5$; $\gamma = 0.5$; $\delta = 0.44$; $F_{drive} = 0.2$; $\omega = 0.735$ units. In this case, the oscillator is characterized by an unstable focus, lying in the negative region of the phase space and this equilibrium point repels the trajectories. Consequently, most of the time the system oscillates only in the right region of the phase space for sufficiently large α values (Fig. 7.5a). Looking at this in the time domain situation (Fig. 7.5b), it is noticeable that the system exhibits small amplitude oscillations (positive instantaneous values only) most of the time and moves to the next potential well intermittently, resulting in a large amplitude (positive and negative values) oscillations. This type of oscillation is corresponding to MMOs. This is because, the

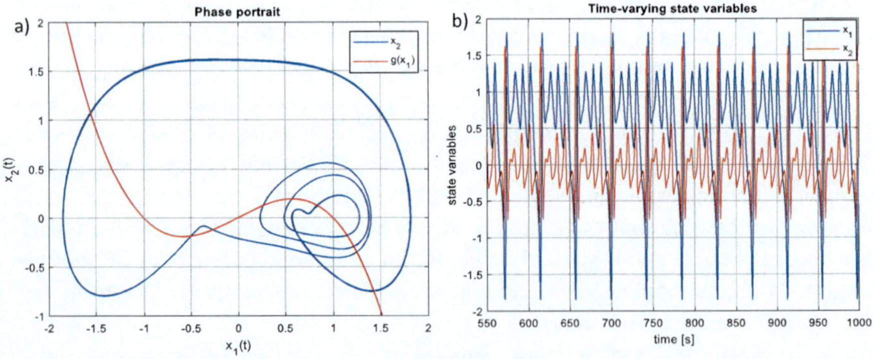

Fig. 7.5 Stable periodic orbits: **a** trajectory and nullcline in the phase space, **b** oscillations of x_1 and x_2

oscillator is characterized by an unstable focus, lying in the negative region of the phase space and this equilibrium point repels the trajectories.

The system exhibits bursting-like oscillations (BOs) for a range of nonlinear damping parameter values. To 1 oscillations, the authors have calculated the total time (T) spent by the system in the right and left potential-wells, namely TR and TL, respectively, and estimated the ratio Tratio = TL/TR. For Tratio > 0.5, the oscillations are characterized as double-well oscillations; otherwise, they are quantified as BOs. If one were to increase α further, the large-amplitude chaotic bursts are found to occur occasionally (with the value of Tratio < 0.1) and randomly with highly unpredictable nature along with small-amplitude chaos. This dynamical state is separately characterized as extreme events (EEs) and the threshold height Hs is used to distinguish it from the other dynamical states. Further, at a critical value of α, the large-amplitude oscillations are suddenly reduced and the system exhibit single-well chaos with the value of Tratio = 0, which then eventually leads to periodic oscillations (via reverse period-doubling bifurcation) for larger values of α. One can note that when α has negative values, then the trajectories are attracted into the left potential-well.

In order to present the essential properties of such a system, it is most convenient to examine first the behavior of a relatively simple autonomous system in the form of an anharmonic oscillator with a cubic nonlinearity. Thus, a pre-tilted, non-vibrating graphene sheet free from any external excitation is examined firstly, which is described by the equations.

$$
\begin{aligned}
\dot{x}_1 &= x_2 \\
\dot{x}_2 &= -0.44x_1x_2 + 0.5x_1 - 0.5x_1^3 + 0.2\sin(0.7366t)
\end{aligned}
\tag{7.8}
$$

As a result of computer simulations (Fig. 7.6), the system undergoes a transition from double-well chaotic oscillations to single-well chaos mediated through a type

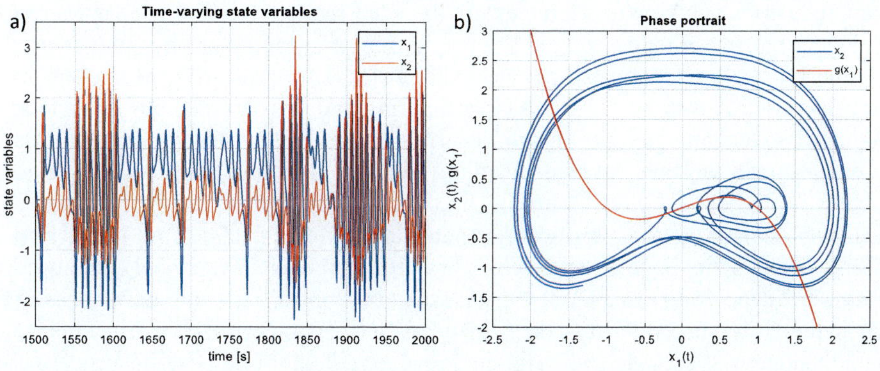

Fig. 7.6 Autonomous graphen system with anharmonic oscillations: **a** steady state variables, **b** phase portrait

of mixed-mode oscillations in which the small-amplitude single-well chaotic oscillations are interrupted by rare and recurrent large-amplitude (double-well) chaotic bursts.

7.2 MMOs in Magnetron Sputtering Processes of Nanomaterials

7.2.1 *Preliminaries*

In surface engineering, magnetron sputtering processes are used to improve the performance of metal products by depositing appropriate metal or dielectric layers on their surface, protecting them against corrosion and wear. In such a process, dimensions from a few atoms to several hundred atoms in the structure of the produced layer should be taken into account. Then, specialized production and research tools are used, including the leading position of X-ray and neutron sources, as well as research facilities at the nanoscale, such as an atomic force microscope, scanning and transmission microscopes, and high-performance computers. They provide unprecedented insight into the structure and dynamics on the atomic scale of materials and on the molecular scale for internal processes. Current efforts in the field focus on optimizing the process of surface layer material synthesis basing on the following factors: plasma ionization state, layer growth kinetics, structure and layer morphology, chemical and phase composition, and on the appropriate sputtering energy supply. This applies mainly to industries such as cosmonautics, aviation, automotive industry, military and police armaments, microelectronics, computers, energy, biotechnology and other related ones. The basis of techniques widely used in production of nano-structured materials and nano-structured surface coatings, which are widely used in many industries applying to a large extent the most modern technologies, appeared in magnetron sputtering processes. The magnetron sputtering method is unique and suitable for the production of specific materials, mainly protective layers, and in the near future it may be more effective than previously used methods in surface modification technologies for known materials, as well as for newly manufactured products. It is a physical vapor deposition (PVD) method and has recently gained impetus in the production of thin films due to its advantages such as greater adhesion to the substrate, faster application speed and uniform thickness on large sized substrates and better deposition stoichiometry compared to conventional techniques of physical deposition of material vapors. One of the most important advantages of this method is its susceptibility to systematic modernization based on the elementary implementation of cathode sputtering in a magnetic field. Moreover, this process is widely regarded as environmentally friendly, as any chemical reagents are avoided during its implementation [3, 4].

Supplying the magnetron sputtering process from a source of current pulses with constant time intervals is the latest achievement in the technology of applying nanocrystalline coatings and has many advantages over others. Modeling of the nucleation mechanism and the material growth mechanism during its magnetron sputtering enables experimentation minimized, especially at the expensive stage of pilot installations, and the cost of obtaining an optimal solution can be significantly reduced. The research carried out so far shows that pulse modulated magnetron sputtering is an effective technique for controlling the thin layer application process in order to obtain unique structures with improved properties compared to coatings obtained by direct current (DC) [9, 10]. This method makes possible to use a higher instantaneous current density during layer deposition, which results in an increased nucleation rate leading to a fine-grained and uniform appearance of the coated surface. The frequency of the power pulses, the duty cycle and the level of the output voltage have a direct influence on the surface properties of the produced layer, its structure and quality, while the stability of the process and its control are improved thanks to the relatively simple system of electric arc prevention and regulation of the relaxation time. Nonlinear interactions between the quantities involved in the magnetron sputtering process can trigger the emergence of highly characteristic oscillatory behaviors known as quasi-periodicity, modal mixed dynamics, and chaos.

When the vacuum chamber is fed with a periodic bipolar current, the voltage between the electrodes varies in time by distinctive fashion. In such a case, as mentioned above, a significant problem is the mixed mode oscillations (MMOs). Due to the interdisciplinary presence of MMOs in the magnetron sputtering process, there is a need to model them in a reliable manner so as to achieve a quick and powerful simulation of state variables in an environment where systems of different nature interact with each other. Overall, modeling of the manufacturing processes of nano-structured materials is extremely complex in theory and has therefore become the field of computer simulation. In the next subsections the dynamics of an anharmonic oscillator with cubic nonlinearity in the presence of linear damping, nonlinear damping, and periodic external forcing will be investigated in more details.

7.2.2 Concise Characterization of the Magnetron Sputtering

An effective inspection of the nature of the mechanisms influencing the magnetron sputtering process and its effects can be obtained by analyzing the operation of the vacuum chamber from the diagram shown in Fig. 7.7. In practice, the magnetron sputtering chambers are realized in structures with different geometrical configurations (cylindrical, planar) and are supplied with energy in different ways (direct current, radio frequency voltage, periodic current with reverse pulse). The schematic diagram of the vacuum chamber in which the magnetron sputtering process takes place is presented in Fig. 7.8. Representative waveforms of direct current and with a reverse current supplying energy for magnetron sputtering are shown schematically in Fig. 7.9. In the first phase of the process, the sprayed material (cathode) called the target and the substrate (anode) are placed into the chamber with high vacuum, and then a noble gas (usually argon) is pumped, which under the influence of an electric field E between the anode and cathode is ionized. Argon as a noble gas is very resistant to chemical reaction in plasma gas mixtures and very often serves as a buffer gas.

The structural and technological uniqueness of DPS power supply consists mainly in adjusting conditions of plasma particles life time by means of modulated pulsed magnetron sputtering.

The ion beam acts on the target surface and because of the phenomena caused by the hitting ions, atoms are ejected from the target surface, which transforms the sprayed material into a gas-like phase. The effect of the positive current pulse feeding the chamber is to attract electrons from the plasma towards the target where they can neutralize the accumulated Ar^+ ions (Figs. 7.7, 7.8 and 7.9).

The duration of the pulse, i.e. pulse width, and the value of the positive pulse are suitable for generating an electron current that can neutralize any Ar^+ ions accumulated on the target and then self-maintaining. Bombardment of the cathode by positive ions (and other plasma components) not only releases secondary electrons, but also the atoms of the cathode material that make up the sputtering material. The cathode material atoms reach the glow discharge plasma and collide with electrons and other plasma elements. The magnetic field applied close to the cathode surface forces the electrons to move in closed loops around the magnetic field lines and they travel a much longer path length than in the case of a glow discharge only without a magnetic field. In a cylindrical magnetic glow discharge, the electric field $\boldsymbol{E}$ has a radial direction and thus the charged particles spiral under the influence of the $\boldsymbol{E} \times \boldsymbol{B}$ field. Therefore, magnetron discharges can operate at much lower pressures (in the order of 1–10 Pa) for the same electric current. The Ar^+ ions are accelerated to the cathode by a strong electric field and sputter target atoms reaching the plasma and can be deposited on the substrate. As a result of the condensation of atoms ejected from the target on the coated element, i.e. on the substrate, a thin layer of deposited material is formed [18, 19]. The natural tendency is to generate a stable plasma with

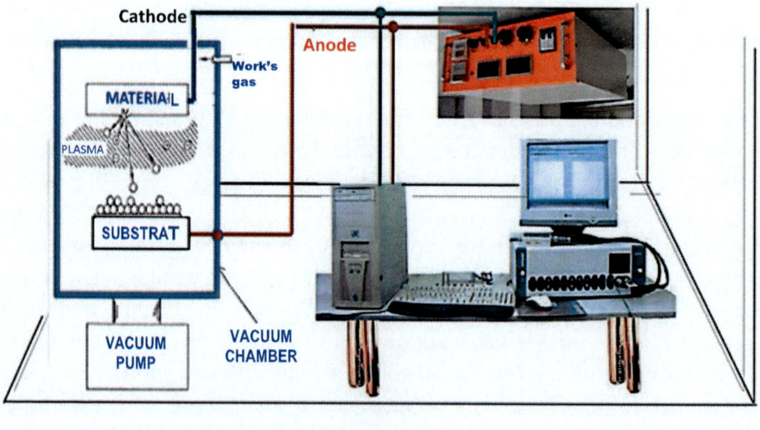

Fig. 7.7 Scheme of the magnetron sputtering appliances

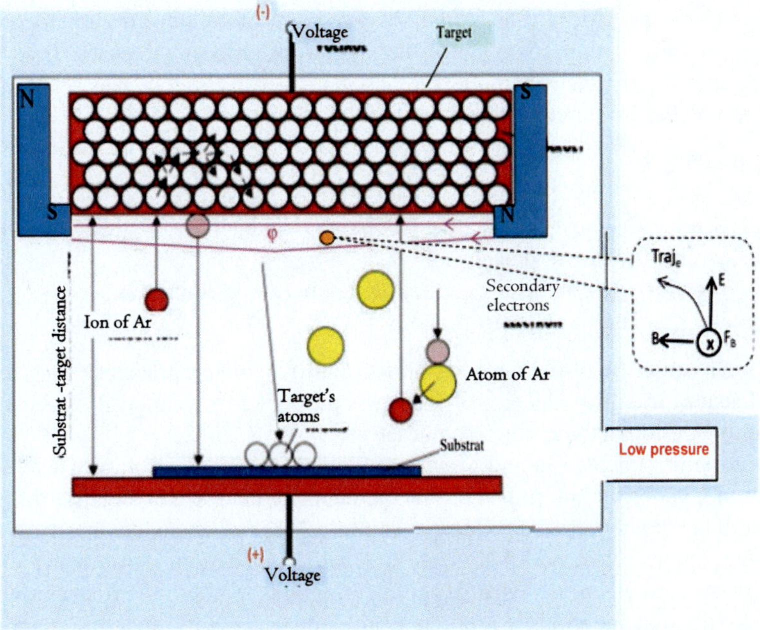

Fig. 7.8 Basic quantities in the process of magnetron sputtering

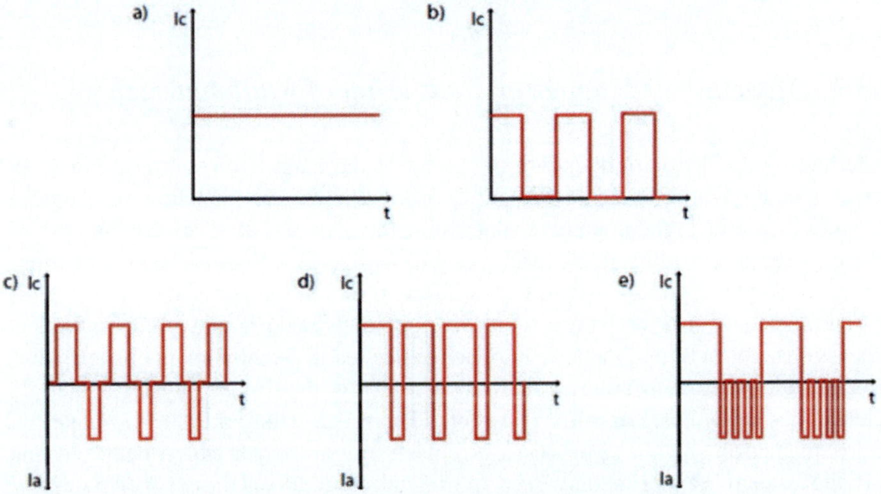

Fig. 7.9 Waveforms of currents supplying the magnetron chamber: **a** direct current (DC), **b** periodic positive pulsating current (PPPC), **c** periodic reverse pulsating current with a dead zone (PRPCDZ), **d** reverse pulsating periodic current (RPPC), **e** pulsating periodic with reversible segments current (PPRSC)

high density of ions. More ions in the plasma mean more ejected atoms from the target, and therefore, the efficiency of the sputtering process increases. It is worth underline that the magnetron sputtering has many advantages over other thin-film application methods, namely:

- high spraying speed,
- easy adjustment of the thickness and structure of the sprayed layer,
- high process efficiency,
- high cleanliness of work pieces,
- improving working conditions and reducing energy consumption,
- no need to regulate gas transformations.

The microstructure of the deposit and its quality, which includes surface roughness, adhesion, impurity, density of the layer produced by sputtering process, is a result of interaction between these parameters.

Condensation, nucleation and growth processes, which in a complex way affect the structure, adhesion and properties of the deposited surface layers, are the result of important characteristics of the magnetron discharge, such as the density profiles of all components, electrode potentials and field distribution, particle and energy fluxes towards the electrodes, profile atomization, gas temperature distribution and energy distribution functions. Significant problems related to modeling magnetron sputtering processes, taking into account the entire main processes accompanying nanostructured products, are presented in the next section.

7.2.3 Modeling of Magnetron Sputtering of Nanomaterials

Determining the factors influencing the effect of the magnetron sputtering process and its variations can be achieved largely by mathematical modeling. The model provides theoretical fundamentals to determine the temporal development process of all the electrical quantities in the discharge gap with respect to the measured external voltage and external total current. It will allow to isolate the contribution of several key parameters from a batch process, trying to predict the dominant effects as well as process changes in time. The basic knowledge needed in establishment of such a tool cover a wide range of disciplines such as plasma physics, surface physics, materials science, gas dynamics, control systems etc. There is a strong drive to simulate the entire magnetron sputtering process, to replace trial-and-error experiments. In this way, the respective experiments can be minimized, especially at the expensive stage of the pilot plant, and then the costs of developing a practical system can be significantly reduced [12–14]. Feeding magnetron sputtering with a current source of reverse pulses in constant time intervals is the latest achievement of nanocrystalline coating technology and has many advantages over others. Accurate modeling of magnetron sputtering as such only makes it possible to understand the course of the process and predict its effects [15–17], and then control it when using this technology in industry. The model becomes more realistic when the input and output are specified.

In terms of control systems theory, the parameters defining the deposition process form the input; the deposited film properties are the output. To improve coatings produced by using that technology the power supplied to the magnetron must work properly and appropriately to desired effect. In the last several years, the experiments have shown that pulsed power supplies provide a much-appreciated service for this area. Currently, high power bipolar pulsed sputtering is gaining interest because of its potential to deposit coatings with unique properties. Under these conditions, extremely high electrical powers are applied in quick succession. Hence, an inherent part of the operation of the magnetron is the effective electric field, which results from an external source of energy and the movement of charged plasma particles. As a result thin deposit properties generally depend on the thickness of the layer, which extends from a few micrometers to nanometer, substrate nature on which the deposits are grown and deposition methodology/conditions used in the manufacturing of thin layers.

Thus, to obtain a complete magnetron sputtering model, it is imperative that it must encompasses the magnetron plasma and its interactions with (some) surrounding solid surfaces. The choice of the proper set of equations is normally based on the given operating conditions. An important point that should be mentioned here is the fact that when a vacuum chamber is supplied with a periodic bipolar pulse current, the voltage between the electrodes can take very different forms of changes in time. Nonlinear interactions between the quantities governing the magnetron sputtering process can trigger the emergence of highly characteristic oscillatory behaviors known as quasi-periodic and modal mixed dynamics abbreviated MMOs as well as chaos. For MMOs, a series of SAOs around the mean value (considered to be "canard" solutions) is subject to a rapid canardial explosion, resulting in the formation of LAOs, which, thanks to a special return mechanism, bring the system close to the mean value. The canard explosion described in detail in [20] is triggered when the trajectory leaves the bend point of cubic nonlinearity ending a series of SAOs and entering into relaxation oscillations with one or more LAOs. An interesting issue with the present problem is that it is very difficult to analyze in detail the situations relating to the nature of charge carriers, the coupling of their motion and the participation of migration in the ion flux. During the ionization and deionization of the flux of gas atoms by current pulses, especially in the case of short pulses, the double layer of the interface between the plasma and the cathode disrupts the pulsed current and affects the overpotential reaction acting on the flux of atoms and ions [10]. The advantage of this technique is mainly expressed in the ease of optimizing the deposition process basing on the following components: plasma ionization state, kinetics of layer growth, structure and morphology of layers as well as chemical and phase composition, etc. To better control the coupled effects of pulsed current supplying the energy to the environment between the electrodes with its own magnetic field and improve the performance of magnetron sputtering, the combined effects must be studied. The Lorentz force acting on the plasma in the between electrode space with higher current densities increases the improving charge flow and conductivity of the ionized gas, which enhances magnetron sputtering performance in terms of accuracy and surface finishing. The azimuth magnetic field associated with the axial flow of a large current through a cylindrical symmetric space with plasma creates a magnetic pressure ($\boldsymbol{J} \times \boldsymbol{B}$ or Lorentz force) that accelerates radially inward of the charged particle flow. For the flat electrodes configuration the system consists of two capacitances namely, C_1 and C_2, where C_1 represents equivalent capacitance of the dielectric region, and C_2 represents the gas region capacitance. As it is evident, with respect to the operating parameters the vacuum chamber can exhibit either filamentary or homogeneous glow discharge regime. The discharge space of the vacuum chamber with the energized plasma exhibits a capacitance C_2 in parallel with a time dynamically varying conductance $g(t)$. The own magnetic field and the heat produced by moving ions and electrons in the chamber are mapped by inductance L connected in series with resistance R.

Taking into account the basic processes occurring both in the vacuum chamber and in the system supplying it with energy, the dynamics of the entire magnetron sputtering system can be described by the following state variable equations where: x_1, x_2, x_3, x_4 denote state variables. The constant parameters a, b, c, d are determined by the properties of the elements making up the system. The minimum value of the conductance of the vacuum chamber is denoted by g_m, and the dot above the variable symbolizes the derivative with respect to time.

The established model predicts the capacitive and Faradic currents of the process, and with the same average and peak current density, the ramp up waveform has higher instantaneous peak current to charge transfer, which results in an improvement in the microstructure of the produced nanocomposite. Due to the complex form of the established system of equations, the determination of their analytical solution is not an easy task and therefore, to examine the effect of various system parameters on the sputtering process, we can perform numerical calculations applying a computer program MATLAB with using effective numerical integration procedures.

7.2.4 Computer Implementation of Model in MATLAB

For over two decades investigations of the operating characteristics of magnetron sputtering processes are intensively carried out by computer modeling and simulation tools. Currently analyzes are beginning to break through to the light of day, showing the memristor properties of a glow discharge plasma in low pressure gases in the presence of an external magnetic field and modally mixed oscillations of the state variables in a magnetron sputtering device [1–9].

In this subsection, the mathematical model of low-pressure discharges in a magnetron sputtering chamber will be used, depending on their physical nature and operation, to explain how it can be integrated into a generalized mathematical structure that reveals memristor-like magnetron sputtering characteristics. It turns out that due to the one-to-one compatibility between the properties of memristors, and the glow discharge plasma, many nonlinear methods of memristor analysis as nonlinear dynamic elements can find useful applications in the analysis of magnetron sputtering nanomaterials. The recent results obtained in this field by many research centers around the world indicate a significant efficiency of the magnetron sputtering process under the conditions of operation with the excitation in the form of reverse pulse periodic current (RPPC) and the voltage response in the form of mixed mode oscillations (MMOs). The voltage-current characteristics of the plasma play an important role in the case of magnetron sputtering. A sequence of small amplitude oscillation (SAOs) around an average value (considered to be "canard" oscillation) is subject to a rapid canard explosion, creating large amplitude oscillations (LAOs) which, thanks to a special return mechanism, bring the system close to the mean value. The canard explosion detailed in [20] is triggered when the trajectory leaves the bend point of the cubic nonlinearity terminating a series of SAOs and converting into relaxation oscillations with one or more LAOs.

Magnetron sputtering works as a result of complex processes of electrochemical reactions, ionization, diffusion of components inside the layer and the vacuum chamber during the deposition operation of the nanocrystalline layers [28, 29]. It is worth noting that all these processes can be assessed by means of memristor current–voltage characteristics. They are mapped by the following memristor properties also known as memristor "fingerprints": (*i*) identical zero-crossing, (*ii*) compressed hysteresis loop, (*iii*) compression of the hysteresis loop depending on the input frequency.

Overall, modeling the sputtering processes of nano-structured materials is theoretically too complex and has therefore become the field of computer simulation. The most popular and powerful software tool for modeling and simulating dynamic systems is Matlab with toolkits such as Simulink and Simpower Systems. It is one of the powerful software packages accepted all over the world.

Typical memristor properties of the tested magnetron sputtering system are visible when i = 75.0*cos(31.40*t)./abs(cos(31.40*t))-50; a = 10,000; b = 10; c = 2000; d = 2500; R = 0.001; g_m= 10^(-8); k = 0.1; I_0 = 400; U_0 = 30,0; P_m= 50. The changes in the supply current and the voltage between the electrodes over time are shown in Fig. 7.10a, while Fig. 7.10b shows the voltage-current characteristic $u(i)$ of the discharge in the vacuum chamber. The plasma conductance trajectory between the electrodes is shown in Fig. 7.10c. The determined waveforms show the formation of MMOs type 5^4 in the system and the typical memristor properties of the studied magnetron sputtering system.

The change in the degree of filling the feeding current period to 25% leads to the MMOs of type 8^{23}, which in numerous cases of magnetron sputtering may prove advantageous due to the reduction of crystalline dimensions, greater hardness of the layer and its better adhesion to the substrate. A further reduction in the degree of filling of the supply current period causes changes significantly different from those presented above, in particular, the system loses its memristor properties, the plasma conduction trajectory takes the form of a single loop and SAOs become high frequency, while the frequency of LAOs also increases, but only slightly. The influence of other system parameters on the effect of the magnetron sputtering process can be determined in an analogous manner. Thus, the designer may have an effective tool to assess the quality of the planned process, its properties and determine the optimal parameters of the elements making up the appropriate system.

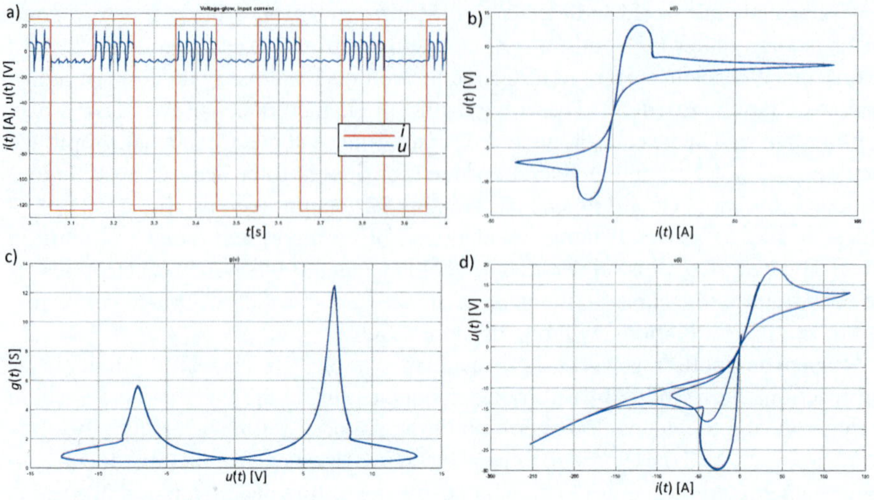

Fig. 7.10 Results of computer simulations: **a** supply current (- - -), voltage between electrodes (- - -), **b** cell voltage-current characteristics $u(i)$, **c** cell conductance trajectory as a function of voltage, **d** cell voltage-current characteristics $u(i)$ after changing only the filling period of the supply current to 25%

The latest achievements of many research groups around the world indicate on a significant efficiency of the magnetron sputtering process under the operating conditions with excitation in the form of periodic reverse pulse current (RPPC) and the voltage response in the form of a mixed mode oscillations (MMOs). This method, with appropriately selected parameters of the supply current, allows generating mixed mode oscillations in the system, which have a positive impact on the effect of sputtering the material from the cathode. This can result in an increased nucleation rate leading to the formation of a fine-grained and uniform appearance of the thin-film coated surfaces. This is of great importance in the case of nanomaterials, the final structure of which is very sensitive to rapid changes in the conditions of the production process. The context of the supply current changes over time is particularly interesting because their influence on the mixed mode oscillations seems to be the most important for the efficiency of the processes. The changing conditions of plasma generation, occurring from the possibility of controlling its frequency and lifetime, have a strong impact on quantity and motion of plasma particles and their interaction with the target and the substrate.

7.3 Dynamics of Train Wheel-Rail Contacts

7.3.1 Introduction

Transport systems are now becoming the basis for integrating vital spheres of human activity in cities and regions, and for their analysis, planning, design and operational functioning in-depth diagnosis of particular system elements are essential. Recent advances in nanoscience and nanotechnology offer particularly exciting possibilities for the development of revolutionary three-dimensional structures that simultaneously optimize components and energy recuperating capacities.

At present, in many research centers around the world are carried out investigations focused on various problems related to the dynamics of rail vehicles. Continuously increasing operational speeds and the demands for comfort require adapting to the top level of knowledge on driving qualities of rail vehicles, safety and comfort of transporting passengers and other loads. In accord to challenges related to higher speed and greater loads with a very high level of security, the use of more and more innovative engineering solutions, better understanding of technical issues and the use of new computer tools are required in the domain of railway vehicle dynamics [22, 25, 27–33]. For instance, identified factors, which have a negative impact on driving comfort, indicate on the need for arranging suspensions that are more advanced.

The basic constitutive element of railway vehicle dynamics is formed by the contact between wheel and rail (Fig. 7.11a). Strictly connected to contact mechanics is the problem of formation of corrugations on the rail treat (Fig. 7.11b), which is now in the centre of attention of many research teams. The knowledge of railway vehicle dynamics allows us to predict with confidence what the values of contact stress, tangential creep forces and creep age in the wheel/rail contact patch are for a

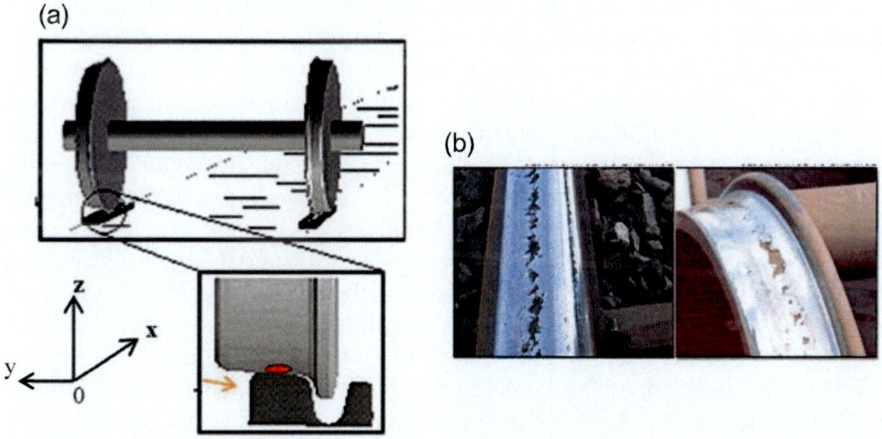

Fig. 7.11 Scheme of the contacting railway parts: **a** rail, wheels, axe and contact frame, **b** spalling from high contact stresses of the rail and wheel

wide range of different conditions. This gives a valuable insight into the influence of the many different factors that affect the incidence of rolling contact fatigue in rails due to the combination of these parameters [17, 20–22, 24, 34–38]. Usually, railway vehicles operating in modern countries use wheel sets comprising two wheels fixed to a common axle (Fig. 7.12). Wheels are rigidly connected with the axle and roll in the direction, which they are heading for. As both wheels are rotating at the same speed, the contact forces ultimately are repartitioned symmetrically on both wheels. These forces are a major cause of the rails wear.

Through mathematical analysis it is possible to build a deep and functional understanding of the wheel-rail interface, suspension and suspension component behavior, simulation and testing of mechanical systems interaction with surrounding infrastructure, noise and vibration.

In its simplest form the equations of dynamic motion of a wheel-rail system are given by a non-linear second order system of ordinary differential equations (ODEs) of moderate dimension that may be solved numerically by standard methods such as incorporated in the packet MATLAB [7–10, 12–16, 18, 19].

7.3.2 The Essence of the Problem

Modeling accompanying the analysis of the processes taking place at the interface between train wheels and the rail surface plays an important role in railway practice. This is due to the systematic efforts during the railways exploitation to reduce the costs of traction maintenance, to effective technical diagnostics of the track and train vehicles, to human safety and the limitation or even complete elimination of negative environmental impacts. During the movement of the train, the contact zone of the bulk materials is perceived as strong enough to withstand the normal (vertical) forces exerted by heavy loads and the dynamic response due to track and wheel unevenness. A hard problem arises when the wheel-rail contact is subject to an action of time-discontinuous forces. The contact zone between a railway wheel and rail is relatively small (roughly 1 cm^2) compared with their overall dimensions and its shape depends not only on the rail and wheel geometry but also on how the wheel meets the rail influence. Thus, studying the complex motions of railway vehicles could give significant new insight into the defect properties of material crystalline structures. It should be emphasized that it is not possible to materials grow without dislocations and/or other disturbances to crystalline order, such as vacancies, interstitials, or substitution impurities. In the case of polycrystalline materials, the memory features of hysteresis may be important according to the methods of their fabrication. Long before defects escalate to the point of incipient failure, they still influence vibrations.

The development of rolling contact fatigue in rails depends on the interplay between crack growth, which is governed by the contact stress and the tangential force at the contact patch. Moreover, the wear depends on the tangential force and the creep age at the contact patch (Fig. 7.11b). These parameters are dependent on a large number of inter-dependent factors, in particular:

- wagon structure: wheel base, axle load, wheel diameter,
- suspension design: in particular primary yaw stiffness,
- track geometric quality,
- wheel profiles: deviation from the nominal profile and state of wear,
- rail profiles: deviation from nominal profile and state of wear,
- wheel/rail friction,
- cant deficiency (depending on speed, radius and cant),
- traction and braking forces,
- wheel and rail material properties.

This exhibits a valuable information on the meanings of the many different factors that govern the range of rolling contact development in rails due to the simultaneous action of the combination of some of these parameters [17, 20–22, 24, 34–38]. Thus for modeling processes in the wheel-rail contact two aspects have to be considered: (1) the geometric and kinematical relations of the wheel-rail contact, and (2) the contact mechanical relations for the calculation of the forces stimulating the wheel and the rail. Hence, an important problem arises with regard to the wagon wheel running on a resilient rail supported by sleepers put flexibly on a completely rigid base. Estimates have shown that in the case of the wheel-rail system, the length of the elastic correlation significantly exceeds the length of the contact area (about 1 cm). The multi-body dynamics and wheel-rail contact models provide invaluable service in this respect, enabling a detailed analysis of not only driving and handling, but also determining their impact on the formation of structural heterogeneity of rails. Experimental studies have shown that the estimation of both the size of the contact area and the pressure distribution in the contact depends on the accuracy with which the surface micro-relief is determined. To overcome this problem we can apply Lagrange's equations to establish equations describing the movement of the wheels while the train is moving. In this way founded the model of the dynamics of the wheel-rail contact, makes it possible to take into account the nonlinear characteristics of the sleeper with respect to the ground. Then the focus is entirely put on modeling the energy absorbed by the rail. In such an approach, the applied method uses energy state variables as time functions leading to the determination of the susceptibility of a given contact to the force generated by train wheel rolling. Under operations of periodic discontinuous forces, the energy absorbed in contact can be measured by loops of one-period energy on the energy phase plane. It is also feasible simulation and testing of mechanical systems, interaction with surrounding infrastructure, noise and vibration. Moreover, through mathematical analysis it is permissible to build a deep and functional understanding of the wheel-rail interface, suspension and suspension component behavior [39, 40].

A large part of the wheel-rail contact modeling leads to the load transfer when touching, and more generally, when there is a force jump between two contact points on the profiles. We assume that the wheel set is rigid and in the rail model, the discrete sleepers under the rail are resting on completely rigid foundation. The interface between the wheel and the rail is a small horizontal contact patch. The contact pressure on this small surface is closer to a stress concentration than in the rest of the bodies. The centre of this surface is also the application point of tangential forces (traction and braking F_x, guiding or parasite forces F_y, see Fig. 7.11a). The knowledge of these forces is necessary to determine the general wheel set equilibrium and its dynamic behavior [41].

For the overall model, all specific phenomena occurring in the system during the train movement on the rail are to be considered. For these purposes, a scheme presented in Fig. 7.12a is adequate. To have a large set of information on the contact, as a function of the vertical relative displacement d_z the study of a single wheel-rail pair is enough. The first modeling of the flange contact is presented to consider it as an elastic spring whose reaction comes from the track and the rail beam deformation. The rail is considered as Timoshenko beam, which can be divided into small segments so that a sleeper is assigned to every segment. The sleeper has both a mass and the pad between the rail and sleeper and the ballast are replaced by springs and dampers. As a natural improvement, a more complex model is also admissible. Consequently, it takes into account the roll effect of the other wheel-rail pair of the wheel set. The

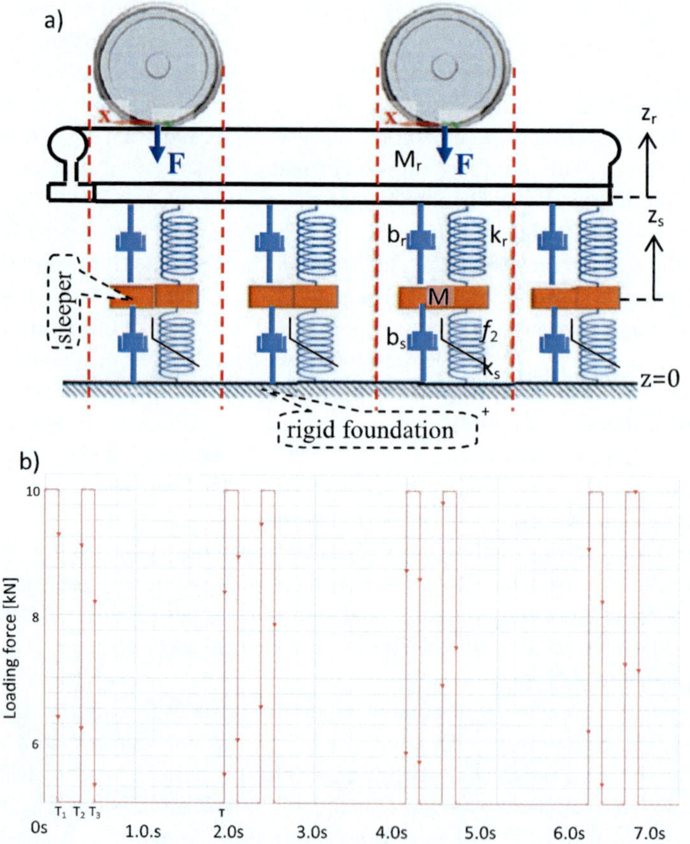

Fig. 7.12 Scheme of: **a** rigid rail with sleepers on the ground, **b** loading force versus time

lateral displacement and the yaw angle can be considered as two small displacements relative to the track and they do not exceed approximately ±1 percentage.

The rail structure is particularly subject to dynamic load, which is induced by moving wheels of the vehicles. Assuming that near the contact patch, the curvatures of the wheel and rail are constant, and then the imprinted contact patch is small compared with the radii of curvature and the dimensions of the wheel and rails. The contacting bodies can be represented by elastic half-spaces and their shape can be approximated by quadratic surfaces. Normally, the material properties of wheel and rail are quite similar and in such a case the tangential tractions do not affect the normal pressures acting between the bodies. Then with these assumptions, for the case where the wheels and rails are smooth, the dimensions of the contact area can be obtained from the theory of Kalker described in [34]. In this case, the contact problem translates to the normal action between the wheel and rail. It means that the normal pressure at a point of the contact patch is proportionate to the interpenetration of the contacting bodies at the point. This approach leads to inertial, stiffness and damping properties varying piecewise-continuously with respect to the spatial location. However, a continuously vibrating system may be approximately modeled by an appropriate set of lumped masses properly interconnected using discrete spring and damper elements (Fig. 7.12). An immediate advantage resulting from such a lumped-parameter representation is that the equations governing the system dynamics become ordinary differential equations [42–45].

For each of segments as that shown in Fig. 7.12a a system of equations can be formulated as follows

$$\begin{bmatrix} M_r & 0 \\ 0 & M_s \end{bmatrix}\begin{bmatrix} \ddot{z}_r \\ \ddot{z}_s \end{bmatrix} + \begin{bmatrix} b_r & -b_r \\ -b_r & b_r + b_s \end{bmatrix}\begin{bmatrix} \dot{z}_r \\ \dot{z}_s \end{bmatrix} + \begin{bmatrix} b_r & -b_r \\ -b_r & b_r + b_s \end{bmatrix}\begin{bmatrix} \dot{z}_r \\ \dot{z}_s \end{bmatrix} + \begin{bmatrix} k_r & -k_r \\ -k_r & k_r \end{bmatrix}\begin{bmatrix} z_r \\ z_s \end{bmatrix} = \begin{bmatrix} F(t) \\ -f_2(z_s) \end{bmatrix} \tag{7.9}$$

Designations in Eq. (7.9) are as given in Fig. 7.12a. The system vibrations are excited by the loading force $F(t)$ exhibiting variations in time shown in Fig. 7.12b. It is worth noticing that the short duration pulses correspond to small dimensions of the contact areas. The duration and intensities of these pulses depend on the vehicle mass and the speed of the train. The supplying T- periodic force $F(t) = F(t + T)$ with two pulses within each period can be represented as follows

$$F(t) = F(t+T) = \begin{cases} A \text{ for } 0 \le \mathrm{t} \le \mathrm{T}_1 \\ 0 \text{ for } \mathrm{T}_1 \le \mathrm{t} \le \mathrm{T}_2 \\ A \text{ for } \mathrm{T}_2 \le \mathrm{t} \le \mathrm{T}_3 \\ 0 \text{ for } \mathrm{T}_3 \le \mathrm{t} \le \mathrm{T} \end{cases} \tag{7.10}$$

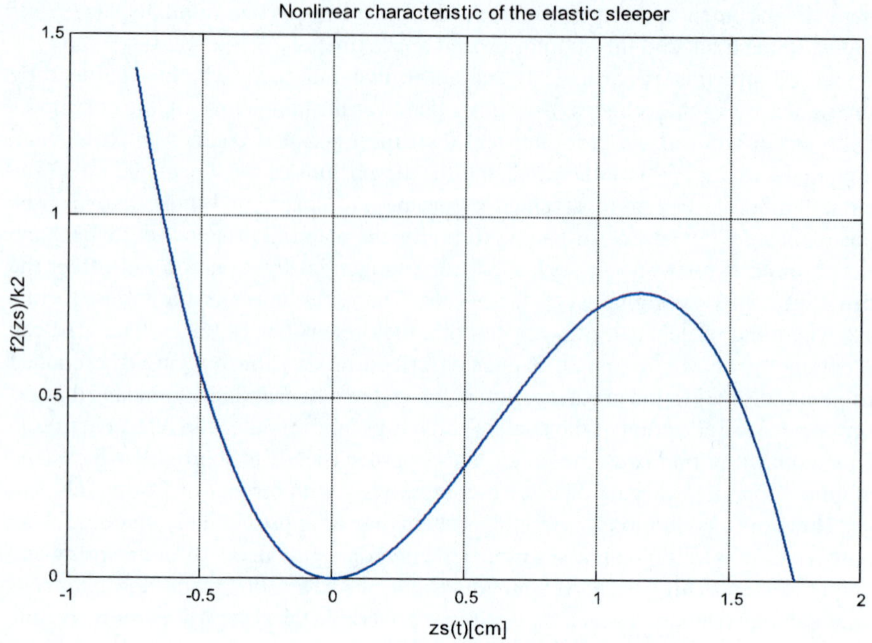

Fig. 7.13 Nonlinear characteristics of the sleeper stiffness

where A and T_k, with $k = 1, 2, 3$, denote the magnitude and moments, respectively, describing the pulses in the supplying force. The duration and intensities of these pulses depend on the vehicle mass and the speed of the train.

The nonlinear characteristic of the sleeper with respect to the ground is represented by the relation

$$f_2(z_s) = k_2 \cdot (\alpha z_s^2 + \beta z_s^3) \tag{7.11}$$

where constant parameters k_2, α and β can be considered as bifurcating values. For $\alpha = 1.75$, $\beta = -1$ the plot of the relative value f_2/k_2 is shown in Fig. 7.13.

In this approach, periodic solutions were determined by appropriate modifications of standard numerical solver used for solutions of nonlinear ordinary differential equations [46]. They appear as computationally not expensive alternatives to the traditional harmonic balance approach and lead to quite satisfactory results. For a set of system parameters and the forcing term shown in Fig. 7.12b and nonlinearity presented in Fig. 7.13 the calculated time variations of the wheel-rail contact deformations are depicted in Fig. 7.14

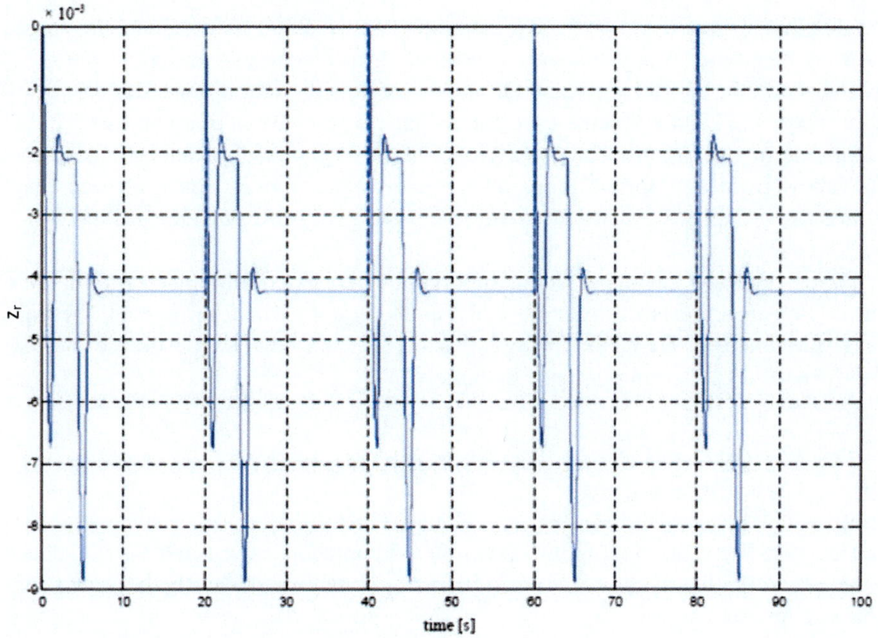

Fig. 7.14 Wheel-rail contact deformations

7.3.3 Computer Simulations

Using modern computer packages it is possible to carry out realistic simulation of the dynamic behaviour of railway vehicles. In the time-domain, integrating (7.9) can always find a periodic solution for (7.9)–(7.11) after the transient responses die out [36, 37]. Such an approach, known as the 'brute force method', is a rather time-consuming task and computationally expensive, particularly for the slowly varying systems. On the other hand the frequency domain approach, known as the harmonic balance technique (HBT), is an iterative method, which matches the frequency components (harmonics) of a set of variables defined for the two sides of (7.9). Although the HBT avoids the computationally expensive process of numerical integration of (7.9), its serious drawback is the large number of unknown variables that must be determined. However, it should be emphasized that a discontinuous signal $F(t)$, like the square wave, cannot be expressed as a sum, even an infinite one, of continuous signals. The extraneous peaks in the square wave's Fourier series never disappear; they occur whenever the signal is discontinuous, and will always be present whenever the signal has jump discontinuities. Thus, it is evident that for accurate analysis of large systems and complicated harmonic producing terms more formal acting time-domain mathematical tool is needed. Taking into account the above requirements and insufficiencies of the methods based on Fourier series, which are up-to-date most commonly used for studies of periodic non-harmonic states of linear systems we propose in the sequel new method for obtaining, in closed form, the response of any system corresponding to piecewise-continuous periodic non-harmonic forcing terms. The method presented in the sequel depends on a "saw-tooth waveform" and a scheme used for unified representation of complex periodic non-harmonic excitations. The main feature of the method is that that its fundamentals follow from appropriate loops on the energy phase plane leading to an easy estimation of the delivery energy to the wheel-rail contact through the evaluation of the loop's area.

Taking into account the periodic supplying force $F(t)$ with pulses within each period we can represent it by more convenient formula when introducing the unit Heaviside functions $H_k(t,T_k)$, $k = 1, 2, 3$ which are shifted at the portions of period with respect to the origin point $t = 0$, namely

$$F(t) = A + \sum_{k=1}^{3} H(t, T_k)[(-1)^k A] \tag{7.12}$$

The periodicity of this imposition can be guaranteed by replacing the variable t with a new variable $p(t)$ called the period carrying waveform or shortly the periodizer, which is defined by

$$p(t) = \frac{T}{2} - \frac{T}{\pi} \arctan(\cot(\frac{\pi}{T}t) \tag{7.13}$$

where T denotes the period. In the presented procedure the solution is constructed for one period [0, T) and then extended to be periodic on the whole t line. This process is called the segment concatenation.

Consider now the elementary segment of the system shown in Fig. 7.12a with an elastic sleeper that is connected with the rigid foundation with a spring and a damper. The data given above are supplemented with the sleeper mass $M_s = 156$ kg and remaining parameters: $b_s = 520$ Ns/m and $k_s = 1.5$ kN/m. Assuming a number of excitation points equal to the number of wheels in one side of the bogie, in the case of twoo wheels gives the input force pulses also equal to two. In such a case the rail deflection takes another form, which is presented in Fig. 7.15a. The corresponding phase portrait $x_3(x_1)$ takes the form shown in Fig. 7.15b.

A very interesting case is that when the wagon wheel presses on a rail in the shape of sawtooth pulses (Fig. 7.16a). Now the response of the wheel-rail system takes a more complex form compared to the case with force in the form of square pulses. During the operation of the force, the response of the system is in the form of high-amplitude bursting vibrations, and after its extinction, low-amplitude SAOs vibrations appear (Fig. 7.16b, c). The situation is different with regard to the sleeper. In the time intervals when the response shows a large amplitude, the bursting vibration is less severe and after its decay, the SAOs amplitude is relatively large. This situation leads to an increase in energy losses in the system and the absorption of more energy by the rail. The presented methodology can be applied to predict the durability of

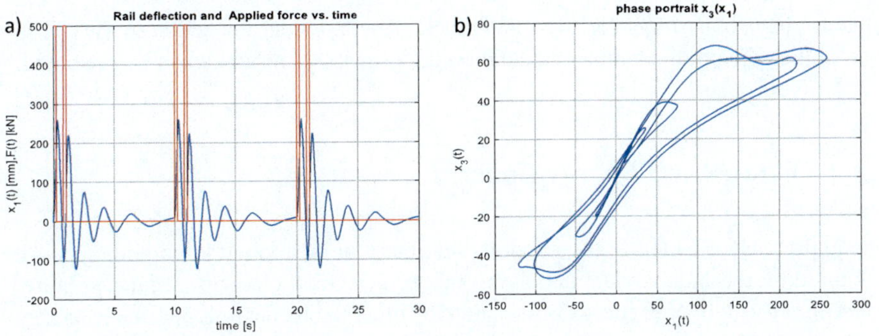

Fig. 7.15 Simulation results in the case T = 20 s: **a** input and output variables versus time, **b** phase portrait $x_3(x_1)$

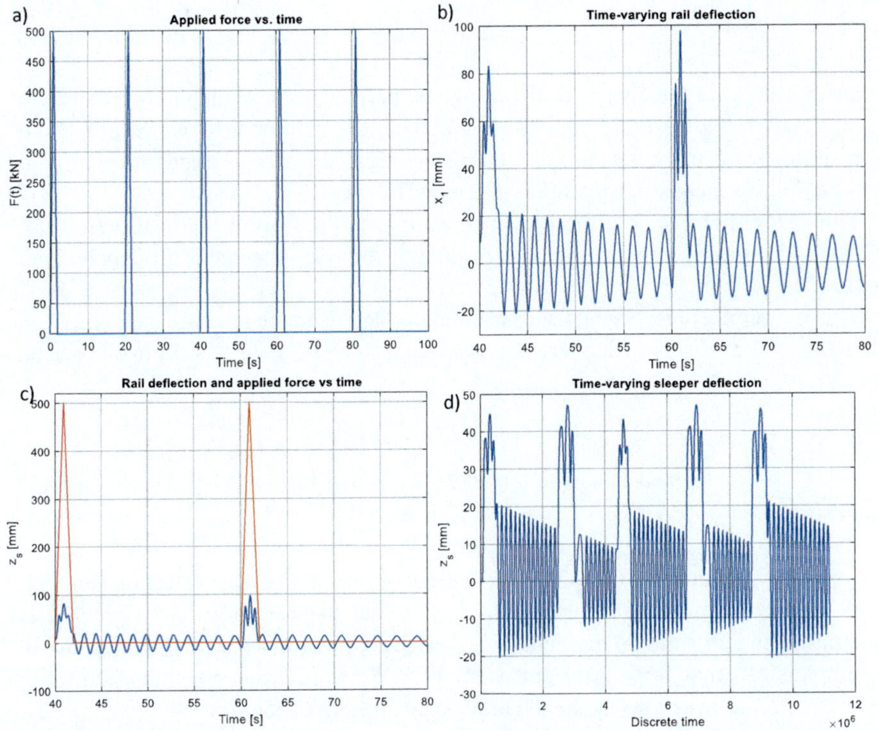

Fig. 7.16 Simulation results: **a** applied force of saw tooth type, **b** rail deflection, **c** rail deflection and applied force, **d** sleeper deflection

wheel rail systems subject to wear and crack growth. Under the action of the cyclic load obtained from the contact calculation the growth of the crack can be predicted.

7.3.4 One-Period Energy Concept

The approach presented in the previous sections can be applied to compute energy delivered by the independent supplying force to the load elements. Under periodic conditions the delivered energy is measured by one period energy loops on the energy phase plane. Consider the steady state energy $W(\Delta t)$ transferred from or delivered to a system element during the time interval $\Delta t = nT$, where $n \gg 0$ denotes a positive integer. It is expressed by

$$W(\Delta t) = nW_T \tag{7.14}$$

where W_T denotes the energy flow during one period of the input and output waveforms. Thus in the periodic non-sinusoidal state it is sufficient to evaluate W_T and then multiplying it by n yields the energy generated (or consumed) by a system element during the given time interval $\Delta t = nT$.

Thus, from here and on, the goal is to establish W_T. In order to deal with it our analysis will follow similar line of argument to the one presented in the precedent section.

This new concept has been discussed in more details in the recent papers [31–33]. Then the total energy, W_T delivered by supplying force $F(t) = F(t+T)$ in one period equals

$$W_T = \int_0^T F(t)v(t)dt = \int_{z(0)}^{z(T)} F(t)dz(t) \tag{7.15}$$

where $v(t) = \dot{z}(t)$ denotes the velocity of the rigid rail oscillations. The above integral is of the Riemann Stieltjes type. This means that the energy of the rigid rail is determined in terms of two mathematically derived quantities $F(t)$ and $v(t)$. It should be emphasized that force $F(t)$ and velocity $v(t)$ are interrelated with each other by means of the equations of the dynamic state and corresponding relationships that define the individual elements. Of special importance is the definition of energy under non-sinusoidal conditions. Fundamental properties of the Riemann–Stieltjes integral in the context of the energy characteristics of dynamical systems give possibilities for introducing new quantities describing system elements from a quite different point of view then all up-to-date used ones [16].

Seeking a power theory for interpreting the energy flow within mechanical systems under non-sinusoidal conditions [24, 37], there is still a lack of agreement between engineers as to which multi-frequency power definitions are to be adopted universally.

The search is still hot for a single theory that will conform to the requirements bulleted above. At first consider as a guide for further analyses the use of Riemann–Stieltjes integral to determine the *rms* values of periodic signals with discontinuities. The idea of effective values, often known as the root-mean-square values, arises from the need to measure the effectiveness of a force or velocity source in delivering energy to a passive load. In general, for any periodic function $f(t) = f(t + T)$, the *rms* value is given by

$$F_{rms} = \sqrt{\frac{1}{T}\int_{0}^{T} f^2(t)dt} \tag{7.16}$$

Frequently there are problems with the use of (7.16) for calculations the energy and especially for the power of system elements in the case of discontinuous signals. In what follows a new approach for the determination of the energy of system elements operating in periodic non-sinusoidal conditions is established. The exact calculations for such problems are indeed possible and accurate. After expressing the voltage $v(t)$ in terms of the physically intrinsic quantity $z(t)$ (spatial variable) by

$$v(t) = \frac{dz(t)}{dt} \tag{7.17}$$

yields

$$V_{rms} = \sqrt{\frac{1}{T}\int_{z(0)}^{z(T)} v(t)dz(t)} = \sqrt{\frac{S_l}{T}} \tag{7.18}$$

The importance of (7.18) lies mainly in the fact that it can easily be mapped on the plane (v, z) in the form of a hysteresis loop that delimits the surface S_l. By referring this surface to the period T and then calculating the square root, we set the desired *rms* value. To show the meaning of the above statement let consider a system

$$\begin{aligned} \varepsilon\dot{x} &= -y - (a + 3bw)x \\ \dot{y} &= \alpha(x = ry - z + e) \\ \dot{z} &= -\beta y \end{aligned} \tag{7.19}$$

$$\dot{w} = x$$

which realizes mixed mode oscillations (MMOs). Taking into account the following parameters and initial conditions: $\epsilon = 0.01$, $r = 2.0$, $\alpha = 1.55$, $\beta = 0.15$, e = -0. 04, a = -0.34, $b = 0.05$ and $x(0) = y(0) = z(0) = w(0) = 0.0$ system generates 2^{11} MMOs. The diagram of the output variable $x(t)$ is shown in Fig. 7.17a.

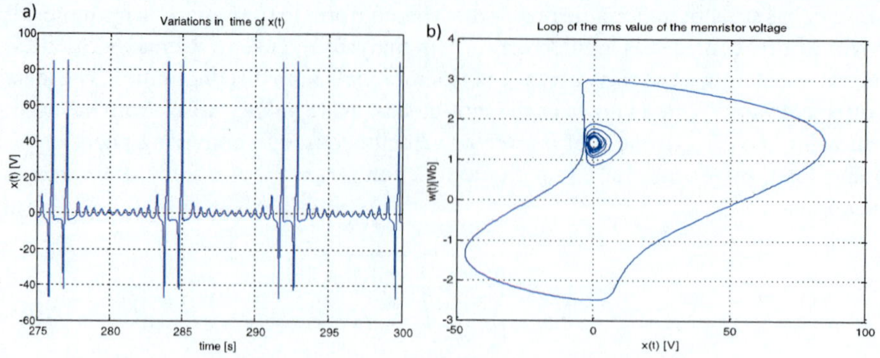

Fig. 7.17 Simulation results: **a** output variable x(t), **b** hysteresis loop for *rms* value

In order to determine the *rms* value of this variable, the use of the expression (7.18) is appropriate, and after representing the integral of the internal expression in the square root on the plane (v, w), we obtain the *rms* histeresis loop, which is shown in Fig. 7.17b. Calculating the surface delimited by this loop applying procedure *trapz* from MATLAB yields $S_l = 392.0512$ m^2/s and in accord with (7.18) we get $X_{rms} = 7.0823$ m/s. The period $T = 7.8160$ s of changes in the instantaneous values of the output variable has been determined using the procedure described in [26] (Fig. 7.17).

Changing initial conditions to $x(0) = 2.22$, $y(0) = -1.105$, $z(0) = -0.00628$, $w(0) = 0.375$ and accepting $a = -0.5$ with unchanged other parameters gives the course $2^4 2^6$ of MMOs shown in Fig. 7.18a. The corresponding hysteresis loop of the output variable *rms* value is presented in Fig. 7.18b. The computed loop's area equals $S_l = 791.0177$ m^2/s and hence $X_{rms} = 8.8939$ m/s.

It is well known, that under non-sinusoidal conditions, some quantities can conduct to wrong interpretations, and others can have no meaning at all. For this reason the need for universally adopted energy identifications has come to the fore

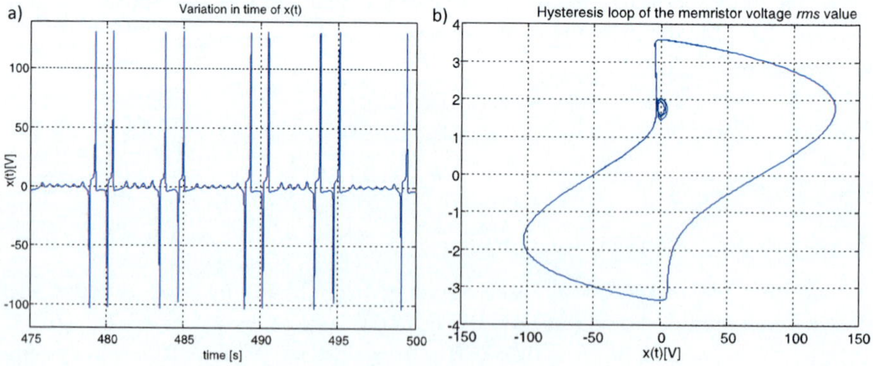

Fig. 7.18. Simulation results ith changing initial conditions **a** diagram of $2^4 2^6$ MMOs of the output variable, **b** loop for the *rms*

now, like never before, in order that designs and energy accounts can be defined and used unambiguously and through which distortion compensation can be specified. Hence, research for a single theory in this area is very hot up to date and should meet the requirements of modern science and technology. Therefore, as mentioned previously, the reliable relationship between the energy of an element and its attributes should be unique and not synthesized. Hence, the energy of the system object should be expressed by two physical attributes, at least one of which should be basic. It follows from expression (7.15) that the area enclosed by a loop on the energy phase plane with coordinates $(f(t), x(t))$ or, equivalently, $(\varphi(t), v(t))$ determines the one-period energy W_T delivered from, or absorbed into, respectively, an object being under periodic non-sinusoidal conditions. Thus for any two-end system component operating in a periodic non-sinusoidal regime it is possible to produce directly the one-period energy W_T without recourse to any forms neither of its power nor the Fourier series approach. The area within the hysteresis loop corresponds to the hysteresis losses, i.e. to the energy dissipated in the load during one cycle of changes of system variables. It is worth mentioning that in the case of conservative system elements their one-period energy loops reduce to a line segment. The solutions obtained by using the periodizer and concatenation procedure can be easily used in (7.15) to determine W_T.

7.3.5 Estimates for One-Period Energy

In the study of dynamical systems arise often problems of determining the range and estimation of a one-period energy of sources or loads for assumed changes in periodic values of non-sinusoidal excitation signals. This problem can be solved by considering the lower and upper limits of the instantaneous values of the corresponding signals associated with the given object. The appropriate estimate can be established by applying the statement of the theorem of Biernacki, Pidek and Ryll-Nardzewski (in brief: theorem of B.P.R.-N.) [47], namely.

If functions $f(x)$ and $g(x)$ with $x \in \Re$ are Riemann–Stieltjes integrable in the interval (a, b) and moreover $m_1 \leq f(x) \leq M_1$ and $m_2 \leq g(x) \leq M_2$, where m_1, M_1, m_2, M_2 denote constants, then the inequality

$$\left| \frac{1}{b-a} \int_a^b f(x) dg(x) - \frac{1}{(b-a)^2} \int_a^b f(x) dx \int_a^b dg(x) \right| \leq \frac{1}{4}(M_1 - m_1)(M_2 - m_2) \tag{7.20}$$

is fulfilled and the constant 1/4 is sharp in the sense that it cannot be replaced by a smaller one. The relatively simple and effective proof of this theorem is given in [47]. Applying the above theorem to the one-period energy expression (7.15) in the case of an electric circuit yields

$$\left| W_T - \frac{1}{T^2} \int_0^T v(t)dt \int_0^T dq(t) \right| \leq \frac{1}{4}(v_{\max} - v_{\min})(q_{\max} - q_{\min}) \tag{7.21}$$

where $v_{max}, v_{min}, q_{max}, q_{min}$ denote maximum and minimum values of the voltage and charge, respectively.

Taking into account the definition of a mean value of a periodic function we get

$$|W_T - V_0 Q_0| \leq \frac{1}{4}(v_{\max} - v_{\min})(q_{\max} - q_{\min}) \tag{7.22}$$

where V_0 and Q_0 denote the mean values of the voltage and charge, respectively.

In the case of a system element exhibiting $V_0 = 0$ or $Q_0 = 0$ we get

$$W_T \leq \frac{1}{4}(v_{\max} - v_{\min})(q_{\max} - q_{\min}) \tag{7.23}$$

Applying the relation (7.23) in the case of an element operating in a sinusoidal circuit with amplitudes of the voltage V_m and the current I_m for a given element, leads to the inequality

$$W_T < \frac{1}{4} 2V_m 2Q_m = TV_m I_m = 2T|S| \tag{7.24}$$

where $|S|=V_{rms}I_{rms}$ denotes the absolute value of the complex power.

This result is consistent with the physical interpretation of the considered problem, since the actual one-period energy of such an element is determined by the expression

$$W_T = TP = TV_{rms}I_{rms}\cos(\alpha) = T|S|\cos(\alpha) \tag{7.25}$$

where P and $cos(\alpha)$ denote the active power and power coefficient of the given element, respectively.

It is important to note that the relation (7.22) in a general case and the relation (7.24) in a particular case allow to estimate the maximal value of the one-period energy of a given two-terminal element on the basis of knowledge of extreme values of its voltage and charge. There is no need to calculate their instantaneous values over the entire period of changes. This eliminates the often-frustrating process of solving a differential equation that describes a given circuit operating in a non-sinusoidal state.

References

1. Trzaska, M., Trzaska, Z.: Nanomaterials Produced by Electrocrystallization Method. In: Aliofkhazraei M., Makhlouf A. (eds) Handbook of Nanoelectrochemistry. Chapt.6, Springer, Cham, 2016, pp.135–168
2. Geim, A.K., Novoselov, K.S.: The rise of graphene. Nat. Mater. **6**(3), 183–191 (2007)
3. Novoselov, K.S., Fal'ko, V.I., Colombo, L., Gellert, P.R., Schwab, M.G., Kim, K.: A roadmap for graphene. Nature **490**, 192–200 (2012)
4. Bogue, R.: Graphene sensors: a review of recent developments. Sens. Rev. **34**(3), 233–238 (2014)
5. Croy, A., Midtvedt, D., Isacsson, A., Kinaret, J.M.: Nonlinear damping in graphene resonators. Phys. Rev. B **86**, 235435 (2012)
6. Chen, C., Rosenblatt, S., Bolotin, K.I., Kalb, W., Kim, P., Kymissis, I., Stormer, H.L., Heinz, T.F., Hone, J.: Performance of monolayer graphene nanomechanical resonators with electrical readout. Nat. Nanotechnol. **4**(12), 861–867 (2009)
7. Kaviyaa, B., Suresha, R., Chandrasekara, V.K., Balachandran, B.: Influence of dissipation on extreme oscillations of a forced anharmonic oscillator. Int. J. Non-Linear Mech. **00**, 1–15 (2020). arXiv.org > nlin > arXiv:2008.13172v1
8. Wicher, B., Chodun, R., Nowakowska-Langier, K., Okrasa, S., Król, K., Minikayev, R., Strzelecki, G., Zdunek, K.: Structure and electrical resistivity dependence of molybdenum thin films deposited by DC modulated pulsed magnetron sputtering. Arch. Metall. Mater. **63**(3), 1339–1344 (2018). https://doi.org/10.24425/123809
9. Trzaska, Z.: Mixed Mode Oscillations in An Electrochemical Reactor. Appl. Chem. Eng. **2** (2019). https://doi.org/10.24294/ace.v2i2.657
10. Kisilowski, J., Knothe, K.: Advanced Railway Vehicle Dynamics. Technico-Scientific Publisher, Warsaw (1991)
11. Trzaska, Z., Marszalek, W.: Periodic solutions of DAEs with applications to dissipative electric circuits. In: Proceedings of the IASTED Conference Modeling, Identification and Control, pp. 309–314. Lanzarote 6–8 (2006)
12. Steenbergen, M.J.: Modeling of wheels and rail discontinuities in dynamic wheel-rail contact analysis. Veh. Syst. Dyn. **44**(10), 763–787 (2006). https://doi.org/10.1080/00423110600648535
13. Santamaria, J., Vadillo, E.G., Gomez, J.: Influence of creep forces on the risk of derailment of railway vehicles. Veh. Syst. Dyn. **47**(6), 721–752 (2009). https://doi.org/10.1080/00423110802368817
14. Iwnicki, S.: Handbook of Railway Vehicle Dynamics. Taylor & Francis, Boca Raton (2006)
15. Nwokah, O.D.I., Hurmuzu, Y.: The Mechanical Systems Design Handbook: Modeling, Measurements and Control. CRC Press, London (2002)
16. de Silva, C.W.: Vibrations. Fundamentals and Practice. CRC Press, London (2000)
17. Bachman, H., et al.: Vibration Problems in Structures. Birkhäuser-Verlag Basel (1995). https://doi.org/10.1007/978-3-0348-9231-5
18. Targosz, J.: Vibroisolation Systems in Railway and Automotive Transportation Applications. AGH University Science & Technology Press, Cracow (2007)
19. Schupp, G.: computational bifurcation analysis: an efficient method for studying stationary and periodic motions. SIMPACK User Meeting Eisenach **9–11**, 1–9 (2004)
20. Benaroua, H.: Mechanical Vibrations. Analysis, Uncertain Control. Marcel Dekker, New York (2004)
21. Schupp, G., Weidemann, C., Mauer, L.: Modeling the contact between wheel and rail within multibody system simulation. Veh. Syst. Dyn. **41**(5), 349–364 (2004). https://doi.org/10.1080/00423110412331300326
22. Feynman, R.P.: There's plenty of room at the bottom: an invitation to enter a new world of physics. Caltech Eng. Sci. **23**(5), 22–36 (1960). http://www.zyvex.com/nanotech/feynman.html

23. Karnopp, D.: Vehicle Stability. Marcel Dekker Inc., New York (2004). https://doi.org/10.1201/9780203913567
24. Zboinski, K., Dusza, M.: Bifurcation approach to the influence of rolling radius modeling and rail inclination on the stability of railway vehicles in a curved track. Veh. Syst. Dyn. **46**(1), 1023–1037 (2008). https://doi.org/10.1080/00423110802037255
25. Blajer, W.: A geometrical interpretation and uniform matrix formulation of multibody system dynamics. Z. Angew. Math. Mech. **81**(4), 247–259 (2001)
26. Marszalek, W., Trzaska, Z.W.: Properties of memristive circuits with mixed-mode oscillations. Electron. Lett. **51**(2), 140–141 (2015)
27. Shabanaa, A.A., Zaazaaa, K.E., Escalonab, J.L., Sany, J.R.: Development of elastic force model for wheel/ rail contact problems. J Sound Vibr **269**(1–2), 295–325 (2004). https://doi.org/10.1016/S0022-460X(03)00074-9
28. Drexler, K.E.: Nanosystems: Molecular Machinery, Manufacturing, and Computation. Wiley, New York (1992)
29. Popov, V.L., et al.: Friction coefficient in 'rail-wheel'- contacts as a function of material and loading parameters. Phys. Mesomech. **5**(3), 113–120 (2002)
30. de Silva, C.W.: Vibration and Shock Handbook. Taylor & Francis, Boca Raton (2005)
31. von Schwerin, R.: Multibody System Simulation: Numerical Methods, Algorithms, and Software. SpringerVerlag, Berlin (1999)
32. Greenwood, D.T.: Advanced Dynamics. CUP, Cambridge (2003)
33. Misiak, J., Stachura, S.: Selected Problems of Static Stability and Dynamics of Rod and Shell Structures. Publishing Office of the Warsaw University of Ecology and Management, Warsaw (2010)
34. Piotrowski, J., Kik, W.: A simplified model of wheel/rail contact mechanics for Non-Hertzian problems and its application in rail vehicle dynamic simulations. Veh. Syst. Dyn. **46**(1–2), 27–48 (2008). https://doi.org/10.1080/00423110701586444
35. Bezin, Y., Iwnicki, S.D., Cavallett, M.: The effect of dynamic rail roll on the wheel-rail contact conditions. Veh. Syst. Dyn. **46**(1), 107–117 (2008). https://doi.org/10.1080/00423110701882348
36. von Wagner, U.: Nonlinear dynamic behavior of a railway wheelset. Veh. Syst. Dyn. **47**(5), 627–640 (2009). https://doi.org/10.1080/00423110802331575
37. Alonso, A., Giménez, J.G.: Non-steady state modeling of wheel-rail contact problem for the dynamic simulation of railway vehicles. Veh. Syst. Dyn. **46**(3), 179–196 (2008). https://doi.org/10.1080/00423110701248011
38. Knothe, K., Stichel, S.: Schienenfahrzeugdynamik. Springer, Berlin (2007)
39. Zerzm, E.: Topics in Multidimensional Linear Systems Theory. Springer, London (2000)
40. Borrelli, R.R., Colleman, C.S.: Differential Equations: A Modeling Perspective. Wiley, New York (2004)
41. Evans, J., Iwnicki, S.: Vehicle Dynamics and the Wheel Rail Interface (2002). http://www.railtechnologyunit.com
42. Trzaska, Z.: Straightforward method for studies of periodic non-harmonic states of linear systems. Arch. Electr. Eng. **53**(2), 191–215 (2004)
43. Trzaska, Z.: A new approach to shaping one-period energies of dynamical systems operating in non-sinusoidal states. Arch. Electr. Eng. **54**(3), 265–287 (2005)
44. Trzaska, Z., Marszalek, W.: Computing periodic solutions of linear differential-algebraic systems with nonsinusoidal excitations. Arch. Electr. Eng. **55**(3–4), 255–271 (2006)
45. Trzaska, Z.: One-period energy in dynamical systems with periodic discontinuous excitations. Elektronika **48**(3), 26–32 (2007)
46. Orlova, A., Boronenko, Y., Scheffel, H., Fröhling, R., Kik, W.: Tuning von Üterwagendrehgestellen Durch Radsatzkopplungen. ZEV-Glasers Annalen **126**(4), 270–282 (2002)
47. Biernacki, M., Pidek, H., Ryll-Nardzewski, C.: Sur une inégalité entre des integrales definies, *Ann.Univ. Mariae Curie-Sklodowska, vol.A* 4, pp. 1–4 (1950)